THE NATURAL HISTORY OF SELBORNE

G+W.
1761.

THE NATURAL HISTORY OF SELBORNE

Gilbert White

Edited with notes by Grant Allen
Illustrated by Edmund H. New

WORDSWORTH CLASSICS

This edition published 1996 by
Wordsworth Editions Limited
Cumberland House, Crib Street
Ware, Hertfordshire SG12 9ET

ISBN 1 85326 181 5

Typeset by Antony Gray
Printed and bound in Great Britain by
Mackays of Chatham, Chatham, Kent

CONTENTS

INTRODUCTION ix

Letters to Thomas Pennant, Esq. 3

Letters to the Hon. Daines Barrington 136

EDITOR'S ADVERTISEMENT 331

Observations on various parts of nature 333

- *Observations on Birds* 335
- *Observations on Quadrupeds* 355
- *Observations on Insect and Vermes* 357
- *Observations on Vegetables* 369
- *Meteorological* 376
- *Summary of the Weather* 380

The Naturalists' Calendars 387

Poems 409

- *The Invitation to Selborne* 411
- *Selborne Hanger – A Winter Piece* 413
- *On the Rainbow* 414
- *A Harvest Scene* 415
- *On the Early and Late Blowing of the Vernal and Autumnal Crocus* 416
- *On the Dark, Still, Dry, Warm Weather, Occasionally Happening in the Winter Months* 416

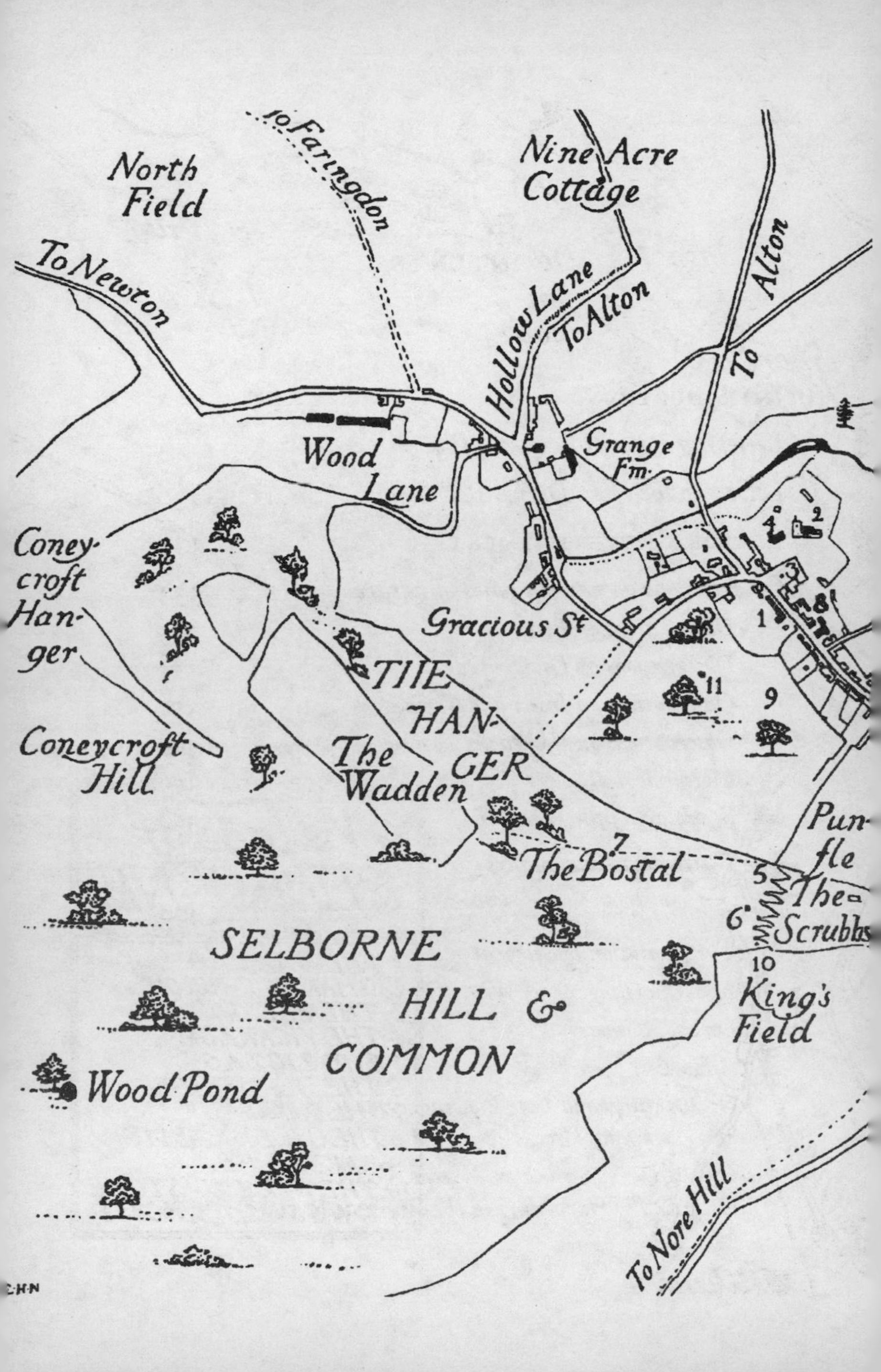
To Faringdon
North Field
Nine Acre Cottage
To Alton
To Newton
Hollow Lane
To Alton
To
Wood Lane
Grange Fm.
Coney-croft Han-ger
Gracious St
THE HAN-GER
Coneycroft Hill
The Wadden
Pun-fle
The Bostal
The Scrubbs
SELBORNE
HILL &
COMMON
King's Field
Wood Pond
To Nore Hill
C·H·N

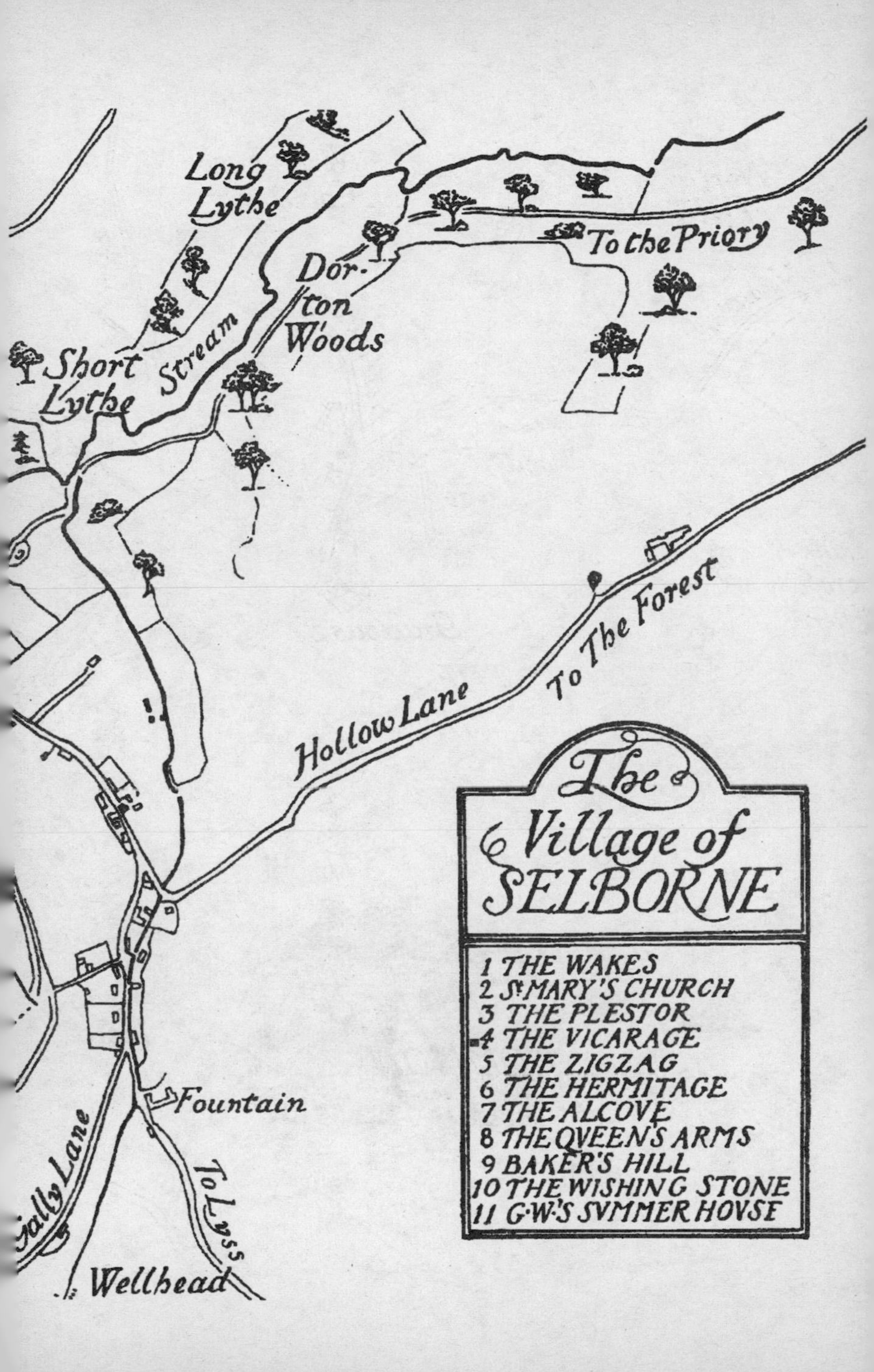
Long
Lythe
Short
Lythe
Stream
Dor-
ton
Woods
To the Priory
To The Forest
Hollow Lane
Fountain
Gally Lane
To Lyss
Wellhead
The
Village of
SELBORNE
1 THE WAKES
2 St MARY'S CHURCH
3 THE PLESTOR
4 THE VICARAGE
5 THE ZIGZAG
6 THE HERMITAGE
7 THE ALCOVE
8 THE QUEEN'S ARMS
9 BAKER'S HILL
10 THE WISHING STONE
11 G·W·'S SVMMER HOVSE

INTRODUCTION

Some time about the year 1755, as far as one can judge, there went to settle down at Selborne in Hampshire a certain quiet and unobtrusive parson, the Reverend Gilbert White, Fellow of Oriel College, Oxford, who has made his out-of-the-way village into a place of pilgrimage for all lovers of nature on both sides of the Atlantic. He was not, as is commonly though erroneously stated, the vicar of the parish; he retained his Fellowship at Oriel, and seems to have resided in Selborne for the most part merely in his character as a private gentleman, though he also incidentally acted as curate there and elsewhere. But that was not Gilbert White's first appearance in the Hampshire parish. He merely settled down to spend his days obscurely but calmly in his native village. So modest was he, indeed, and so careless of fame that no portrait now exists of him, and only a few particulars can with difficulty be gleaned from very brief notices about the man whose Letters have probably been reprinted in a greater number of editions than those of any other English worthy.

The Whites had a hereditary connection of two generations with Selborne. Gilbert White the elder, grandfather of the naturalist, was a Fellow of Magdalen, presented by his college in 1681 to their vicarage of Selborne, then, it would appear, of very small value. The tombstone of this elder Gilbert, still remaining in the parish church, is partly answerable for the persistent blunder which describes the naturalist as 'the idyllic vicar of Selborne'; and the error is intensified by the memorial slab to the grandson himself, on which occur the words, 'formerly Vicar of this Parish', applied to the elder not to the younger Gilbert. The vicar died in 1727, when his famous grandson was just seven years old; he left one son, John, 'Barrister at-law', who was the father of the more famous Gilbert White of these charming letters.

Gilbert the younger was born at Selborne vicarage on July 18,

1720. He died in 1793, so that his seventy-three years fairly covered the greater part of the eighteenth century and of the reigns of the three Georges. Selborne is even now a remote country village, far from the railway; it was then yet more inaccessible and sequestered than at present. It lay midway between two great coach roads, the Portsmouth and the Winchester; and it was approached only by those deep, steep, and water-worn lanes of which White speaks so feelingly, but to render which passable his grandfather the vicar had left a considerable sum of money. Roughly speaking, indeed, Gilbert White spent most of his life at Selborne; and it is partly that long ancestral connection with a single spot which imparts so much value to his continued series of local observations. But he did not lack polite learning, nor intercourse with the best of his kind elsewhere. He went to school at Basingstoke, with Thomas Warton, a well-known clergyman, famous as the father of two more distinguished sons, Joseph, Master of Winchester College, and Thomas, Professor of Poetry at Oxford. Thence the lad proceeded in due course to the University, where he matriculated at Oriel in 1739, being then nineteen. Four years later, in 1743, he took his degree of Bachelor of Arts, and in March 1744, he was elected to a Fellowship. He seems to have resided for at least three years afterwards at the University. His first curacy was at Swarraton, near Old Alresford. In 1752, however, he was Junior Proctor at Oxford, and there are indications that the Swarraton curacy was little more than a title. Not very long afterwards he retired to Selborne, where he was finally settled in 1755, though he did not inherit the family property till his uncle's death in 1763. He could never thenceforward be induced permanently to quit this his chosen place of residence. He refused more than one offer of a college living, preferring his modest curacy at Faringdon, and the quiet ease of a lettered naturalist to parochial distractions. But it would seem from a curious passage in the *Antiquities of Selborne* that before settling in Hampshire he must have passed some time practically as a gentleman farmer in the Isle of Ely.

Such are the chief dates in the Hampshire parson's simple annals. White has left for us an account of his life, however, far more graphic and valuable than any mere formal biography – an account which more than makes up for the want of details as to his external history. That history, as his nephew wrote, 'passed tranquil and serene, with scarcely any other vicissitudes than those of the seasons'. Some time about the year 1767, as it chanced, he entered into a brisk correspondence with Thomas Pennant, a wealthy

Welsh naturalist, the author of the *British Zoology*, with regard to the habits of certain birds and animals. These letters, one may conjecture, were begun without any thought of ultimate publication; the earliest in date among them seem to be written offhand, without order or method, as mere rough memoranda of facts and observations. Letter 10, as at present arranged, is probably the first that really passed through the post between the two naturalists; one may gather from it that Pennant had asked a few questions of White, and that White replied to them seriatim, in the order in which they were written. From this casual beginning, a regular correspondence ensued, carried on for a long time without thought of publication. But gradually another of his correspondents, the Hon. Daines Barrington, seems to have suggested that so much valuable matter ought not to be locked up in private letters; and thenceforth White would appear to have begun aiming at a more regular style and at something approaching orderly treatment. A letter to Pennant in 1771 indicates the probability that the Welsh naturalist too had urged him to publish. Unless I mistake, it is possible to detect in the letters as they proceed, at least from this point, a gradual development towards literary treatment. The series addressed to the Hon. Daines Barrington, another naturalist of the day, began a little later than that already alluded to: its course was in large part contemporaneous with the course of the letters written to Pennant. A similar change of tone may be observed in this correspondence also between the earlier and the later members of the series.

About the year 1784, when France and America were in ferment, White must have finally adopted the design of publishing both sets of letters, though as early as 1776 he had debated the matter. At that time, I conjecture, he produced most of the first nine (artificial) letters as at present arranged; but one of them may perhaps be made up of fragments from a real communication addressed to Pennant. These introductory chapters are not, as a matter of fact, real letters at all: they consist of a general and formal description of Selborne, its position, soil, surroundings, and so forth; and their contents might far better have been thrown into the natural form of a preface, save that White in his excessive modesty perhaps shrank from even the bare appearance of deliberate authorship. A single sentence in Letter 9, however, where he refers to the spring of 1784, allows us to see that these introductory epistles, which pretend to usher in the series, were really an afterthought, designed to put the reader in a position for understanding the matter that follows. One or two subsequent letters, more comprehensive in scheme, I believe to have been added or expanded at the same period.

The first edition of the collected correspondence appeared in 1789, the momentous year of the great French Revolution. It was published by Gilbert's brother Benjamin, a bookseller in London – the same who gave to Selborne Church the beautiful old German altar-piece which still adorns it. White only outlived the appearance of his work by four years; he died in 1793, the culminating year of the terror in Paris. These rough indications of contemporary events will suffice to place him in his own century.

The quiet parson naturalist himself, who thus lived and worked at Selborne, could little have suspected the immense popularity which his work was to attain. In order to understand the peculiar fascination of these sketches and observations, which have passed through edition after edition with increasing frequency, we must consider the special combination of circumstances under which they were begotten.

White's period of literary and scientific activity corresponds roughly with that part of the reign of George III which precedes the French Revolution – say, in brief, the age of William Pitt the elder. Now, intellectually, this was an age of steady though slow progress in England. The general European scientific movement had gathered head in Britain with the establishment of the Royal Society by Charles II. The later seventeenth and early eighteenth century saw a gradual increase in the interest of learned men in natural phenomena, and particularly in the life of plants and animals. The fauna and flora of Europe then first began to be accurately investigated; travel in Asia and America brought knowledge of new forms to the ken of acute European naturalists. Zoology and botany formed just at that date, indeed, what one may venture to call the growing-point of science as astronomy had formed it in the age of Copernicus, and as geology formed it in the age of Lyell. The publication of Linnaeus's great work on *The System of Nature* gave an impulse to the study of biology, the effect of which can scarcely be overrated. It was during the forty years roughly covered by White's observations that the science of life began to assume philosophical form and to be prosecuted with some attempt at scientific accuracy.

Gilbert White, Fellow of Oriel, was a man of highly competent education, a good classical scholar, capable of reading with ease the Latin works and memoirs in which the scientific writing of the time was almost all contained. His *Antiquities of Selborne* show him to have been also a man of great general erudition, with a knowledge of and interest in medieval civilisation very rare in his day. But he was also by nature and habit a keen observer of the wild life around him. When he settled down at Selborne, to a placid bachelor existence, he occupied a house in the main street of the village, still standing, though much enlarged, known as the Wakes; and, being a celibate Fellow, with few cares to worry him, he gave himself up almost entirely to his favourite fad of watching the beasts and birds of his native country. At the present day, unless one devotes oneself to the minuter forms of life, one has little chance of discovering anything new in Britain. But in White's day things were different. The zoology and botany of the British Isles were as yet very imperfectly understood; the habits and ways of plants and animals were an almost unknown study. Moreover, the current books on natural history were still crammed with medieval fables, marvellous survivals of folktales, extraordinary accounts of how swallows hibernate under water, and how decoctions of toads are a certain cure for the ravages of cancer. It was the business of White's generation to substitute careful and accurate first-hand observation for the vague descriptions, the false surmises, and the wild traditional tales of earlier authors.

This it is in great part that gives their perennial charm to these natural, personal, and delightful Letters. We are present, as it were, at the birth of zoology; we are admitted to see science in the making. Europe at that period was full of patient and honest observers like White, on whose basis the vast superstructures of Cuvier, Owen, and later of Darwin, were at last to be raised. But most of them are not, as individuals, forgotten, because they did not personally commit their work to print and paper, save in the transactions of learned societies. In White's *Selborne*, on the other hand, we have crystallised and preserved for us the very stages by which each plane of truth was slowly arrived at. We assist at the deliberations of the early biologists. We see them comparing and identifying species; we find them fighting for or against some hoary but untenable tradition; we note their eager love of truth, their burning desire for exact knowledge, their occasional reluctance to abandon some cherished fable which now seems to us too childish for such men's serious consideration. It is therefore as a historic document that *The Natural History of Selborne* most of all appeals to us; it shows us by what steps science felt its way in the later years of the eighteenth century.

Moreover, it is essential to insist upon the point that the interest of these Letters is now chiefly literary. No other work of science of that age survives practically today. The contents and results of such works, it is true, survive in modern books, so far as they have stood the test of time; but the works themselves are as dead as Scopoli and Linnaeus. Why is this? Simply because science is always growing; and even the best of scientific books become rapidly antiquated. Nobody who seriously wishes today to learn anything about beasts or birds, about plants or flowers, about rocks or fossils, about the laws of nature, would dream of going for facts and observations to authors of the eighteenth century. All that those authors had to say of importance has been adopted, adapted, modified, codified, added to, made more accurate by writers of the nineteenth. When we return upon our steps to read a systematic scientific work of the last century it is never for the sake of its value as instruction, but solely for the sake of its place as a stepping-stone in the history of science.

Letters like White's, however, stand on a somewhat different footing. We read them partly indeed for this same purpose, as moments in the development of biological thought, but still more as vivid and graphic pictures of a phase of existence. Fully to understand *The Natural History of Selborne*, one ought to visit Selborne itself. There, facing the chief street of the village, you see a quiet and unobtrusive old house, which is the one where White made his immortal observations and penned in peace his immortal letters. As you look at the front towards the street, indeed, you wonder that such a site could afford the bachelor parson sufficient opportunities for watching the intimate life of birds and beasts as his correspondence shows him to have watched it. But if you obtain the courteous permission of the present proprietor to enter the house and inspect the garden, you will no longer feel surprised. The front windows, it is true, give upon a very compact street of eighteenth-century domestic architecture; the back opens out upon a spacious lawn and garden, sloping up towards the Hanger, and wooded with fine old trees, some of them doubtless of Gilbert White's own planting. Here the easy-minded Fellow of Oriel and curate of Faringdon could sit in his rustic chair all day long, and observe the birds and beasts as they dropped in to visit him. The Letters are the vivid picture of a life so passed – the life of a quiet, well-to-do, comparatively unoccupied gentleman of cultivated manners and scientific tastes, studying nature at his ease in his own domain, untroubled by trains, by telegrams, by duns, by domestic worries; amply satisfied to give up ten years of his life to settling some question of ornithological detail, and well pleased if in the end his conclusions are fortunate enough to meet the approval of the learned Mr Pennant or the ingenious Mr Barrington.

Those times have passed away. Science has become a matter of special education. The field of the amateur has been sadly curtailed. No man now can hope to attain to new facts or generalisations without the copious aid of libraries, instruments, collection, co-operation, long specialist training. But the calm picture of this more peaceful and easy-going past is all the more pleasant to us on that account. I confess I can never read a page or two of White without recalling to my mind those exquisite lines of Dustin Dobson's which sum up for us the ideal eighteenth-century gentleman:

> *He liked the well-wheel's creaking tongue –*
> *He liked the thrush that stopped and sung –*
> *He liked the drone of flies among*
> *His netted peaches;*
> *He liked to watch the sunlight fall*
> *Athwart his ivied orchard wall;*
> *Or pause to catch the cuckoo's call*
> *Beyond the beeches.*

Such of a surety was Gilbert White's ideal; and we may almost add of him, in Mr Dobson's apt phrase, 'His name was Leisure.' Time was not then money; it was opportunity for enjoyment, for self-development, for culture. And as such White used it, with a consciousness of dignity and a sense of worthiness in life which have almost faded out of our hurried modern existence.

'Tis as a literary monument, therefore, I hold, that we ought above all things to regard these rambling and amiable Letters. They enshrine for us in miniature the daily life of an amateur naturalist in the days when the positions of parson, sportsman, country gentleman, and man of science were not yet incongruous. And in this spirit and from this point of view I have thought it best to edit White's charming volume. I have not attempted the impossible task of bringing our author's biology 'up to date', as a matter of technical modern information. To do so would be to overload the work with useless notes, which could only distract the attention of the reader from what is central and essential to the time and place of the original writer. When White wrote, the very convenient Linnaean system of nomenclature, for example, which marks genus uniformly by one substantive and species by one epithet, had not yet fully superseded the clumsier old descriptive method; so that White frequently refers to birds or mammals by the cumbrous and uncertain many-worded names bestowed upon them by Ray and other early naturalists. I have not in every case endeavoured to correlate these with the accepted modern scientific titles, partly because the identification is often doubtful, but still more because the book must be read in the historic and not in the strictly scientific spirit. You must think yourself back mentally into White's position. On the other hand, I have desired to prevent the work from giving currency to really false or exploded views, and still more from being a source of erroneous ideas as to fact, by correcting in the footnotes (for the benefit of young or untechnical readers) the most questionable or mistaken statements and conclusions. Wherever modern science has authoritatively settled some point which was a moot one for White, I have given its decision without its reasons. Wherever it has pronounced with a clear voice against his speculations, I have briefly chronicled its new view. Where possible with absolute certainty, I have substituted accepted modern names; and I have also brought White's crude local geological nomenclature into line with the terms of modern geologists. I have corrected and emended the text where it was clearly faulty; I have given the more recognised modern names of villages and hamlets in square brackets, while preserving in the text White's spelling; and I have occasionally added the modern form or equivalent of a word which White uses in an obsolete shape or a provincial sense. I have thus confined my work in the strictest sense to the task of editing a classic; I have not attempted the impossible labour of bringing all its statements up to the modern standard of scientific knowledge. And that no doubt may exist as to what part is the author's and what the editor's, I have enclosed all my own additions in the text in square brackets. To my own notes I have added the abbreviation ED. Notes without this addition are therefore those of the original writer.

While saying all this, I would not wish in any way to detract from the solid and permanent scientific value of White's remarkable life-work. On the contrary, it is impossible not to attach the highest importance to it. Most of his observations were conducted with such care and accuracy that they are still among the best we possess for the fauna and especially for the birds of Great Britain. Only a few modern observers, such as Mr Warde Fowler and Mr Hudson, can be named in the same rank with White as patient and sympathetic first-hand watchers of the wild life of the moors and woodlands. Whoever reads these Letters today may learn on every page of them numerous facts which no subsequent observation has either disproved or improved upon. I have lived myself for some years in White's own country, looking out daily upon Selborne and upon Wolmer Forest; the same ponds have flashed in the sunlight on my eyes; the same beasts and birds and insects have darted before me. I have constantly read White's accounts of their habits and manners; and I have been every day more impressed by the depth and width of his knowledge, the accuracy of his observation, the candour of his mind, and the intimate acquaintance he possessed with the outer life of nature in England.

From this point of view, the value of White's work is universal and permanent. His method is even more important than his results. He teaches one how to observe; he shows us by an object-lesson of patience and watchfulness how we ought to proceed in the investigation of nature. In his time, all the work was still to do. In ours, for Europe at least, the greater part of it has been already done. Today, if a boy or a man wants to know about the plants, the birds, the fish, or the insects of the country in which he lives, he usually begins by 'buying a book about them'. He collects specimens, of course, and identifies them with his book; but as soon as he has found out to what particular species each specimen belongs, he generally contents himself with reading up what his book says about it, and then rests satisfied that he has fairly 'done' that plant or animal. Thus the very perfection at which our text-books have arrived stands in the way of first-hand observation. Book-knowledge tends more and more to supersede direct contact with nature. But White may suggest to us a more excellent way. The record of his long years spent in finding out for himself what the beasts and birds really did do makes us feel that books are of little use beside direct eyesight. Nowadays, the traveller in relatively new lands has to watch the fauna and flora as White watched them in England; but at home in Europe it is too often the case that intimacy with printed pages is substituted for intimacy with the objects they describe for us.

Nor is this all. White has another and a higher side. He represents the dawn of the philosophic spirit in science. In no small degree, he leads up to the generation of colossal thinkers – the generation of Lyell, Darwin, Spencer and Huxley.

The learned men of the sixteenth century, it often seems to me, were individually wasted for the sake of humanity that came after them. They spent their lives in useless wrangling over petty points of Ciceronian Latin and Periclean Greek; they accumulated stores of minute learning for which they could suggest no possible employment. But the materials they collected proved useful in time for the evolution of that higher type of scholarship which came out in Gibbon and the French Encyclopaedists, and which has revolutionised the conceptions of ancient literature and ancient history in our own day. These men were like brickmakers who blindly fashion bricks which some great architect may afterwards pile up with broad design into some noble fabric. Even so, I feel, the men of science of the eighteenth century were individually wasted for the sake of the future of their subjects. They collected great masses of unrelated facts, which seem tediously monotonous and destitute of wide informing principles to a modern reader. They wrangled over the identity or distinctness of species. They framed with care endless artificial systems of classification. They noted petty points of structure, apart from function. And for the most part, they did it all without one glimmer of generalisation, one passing glimpse of an idea or a theory. We would think their work impossible did we not know it to be true, and did we not see the same type of mind represented now in the restricted local botanist and ornithologist of today – the man who revels in the splitting of critical species, who discovers some new spot on a butterfly's wing, and who makes it his highest glory to have given his own name to this or that insignificant variety of the common stitchwort or the ordinary earwig.

Gilbert White was one of the few eighteenth-century naturalists who struck the keynote of a higher conception of biology. He was in many ways the forerunner of Darwin and of Müller. His work stands out among the work of his time as conspicuous for its philosophical tone and spirit. He is always observing just those points about life which were afterwards to supply clues to the inner secrets of nature. Thus he notes how the young of the stone-curlew love to skulk among the stones in a flinty field, 'which are their best security; for their feathers are so exactly of the colour of our grey spotted flints, that the most exact observer, unless he catches the eye of the young bird, may be eluded'. This is the germ of the theory of Protective Mimicry. In the same way, his remarks on the influence of food upon colour in Letter 15 to Pennant; his notes on the habits of the swift in Letter 22 to Barrington; and many other similar remarks, show premonitions of the final development of rational biology. As to his prescient observations on the part played by earthworms in the economy of nature, I have already called attention, in my little book *Charles Darwin*, to the extraordinary way in which they anticipate our great biologist's theories and experiments in that direction. Indeed, throughout, White was one of the few early naturalists who recognised the importance of the cumulative effect of infinitesimal factors – a truth on which almost the whole of modern biology and geology are built up. As zoologist, as botanist, as meteorologist, as sociologist, he is possessed in anticipation by the modern spirit in every direction. In this respect, it is true, he cannot be named beside his far abler contemporary, Erasmus Darwin; yet while Erasmus Darwin has left behind him great speculations, immensely interesting to the historian of science and philosophy, but not to the general reader, Gilbert White has produced a book which will continue to be read for years, both as a model of observation, and as the picture of a man, a place, and an epoch.

For White is essentially lovable. We know him as a crony. We can chat with him still, on the slopes of the Hanger, up which he cut the walk still known as the Bostal, about the number of British species of willow-wren, the reason for the separation between the sexes of the chaffinch in winter, and the way to worm out field-crickets from their holes by the gentle persuasion of a bent of grass-flowers. It is the almost colloquial form of the Letters that gives us this sense of nearness and familiarity. Hardly anywhere else are we transported so frankly into the inner atmosphere of the eighteenth century; even Boswell's *Johnson* fails in some respects to come up to the level of this unconscious self-revelation of the gentle, inquisitive, garrulous country parson. We see him traversing on his cob 'that chain of majestic mountains', the Sussex Downs; we hear him speak with bated breath of the awful heights of Snowdon and Plinlimmon; we smile at his naïve allusions to Spain as a distant and almost unknown kingdom; we are amused at the curious restrictions of space which are implied in almost all his references to countries other than European, or even to the remoter parts of Europe. Yet the charm of the picture never once diminishes. Indeed, it is just these quaint touches of vanished thought that make the book most readable. 'I return you thanks for your account of Cressi Hall; but recollect, not without regret, that in June 1746 I was visiting for a week together at Spalding, without ever being told that such a curiosity was just at hand.' Murray and Baedeker were then unknown. Nowadays we should say, 'I will run down to Lincolnshire and look at it': but Lincolnshire to White was further off for all practical purposes than Moscow or Morocco to the modern investigator. This steady picture of a calm and contemplative rural life is worth a thousand times more than much minor science.

Finally, we have always to bear in mind the end which thinkers of

White's age proposed to themselves. In our own day, the desire to 'advance science' has been made on the whole a foolish fetish. Almost all scientific education has aimed at this end; it has striven to produce, not whole and many-sided men and women, but inventors, discoverers, producers of new chemical compounds, investigators of new and petty peculiarities in the economy of the greenfly that affects roses. All that is very excellent in its way; but it is not the sole, or even (let me be frank) the main object of a scientific education. What the world needs is not so much advancers of science as a vast mass of well-instructed citizens, who can judge of all subjects alike in their proper place, and can assign to each its due relative importance. I know few things more instructive in this way than to turn from the *Natural History* to the *Antiquities of Selborne*, and see how far White differed in the width and universality of his broader interests from the narrow and specialised man of science of today. The truth is, the vast majority of men can never do anything to 'advance science' in any noteworthy degree; and the desire to 'fake up' a petty name by pretending to advance it lies at the root of much of our current pedantry. But everybody can love and observe nature. Everybody can take lessons from White in such love and observation. The aim we should propose is to build ourselves up in the round; to make of ourselves full, evenly balanced, broad-minded human natures. We do not want to be lopsided. As a preservative against one prevalent form of lopsidedness in modern life, White's methods and example are of incalculable value. Try to look out upon Nature with the same frank, unprejudiced, first-hand view, asking her questions, and letting her answer them herself, instead of forcing a hasty answer upon her; and then, whether you succeed in 'advancing science' or not, you will at least have advanced our common humanity by the presence in its midst of one more candid and single-hearted lover of truth and beauty.

GRANT ALLEN

THE NATURAL HISTORY OF SELBORNE

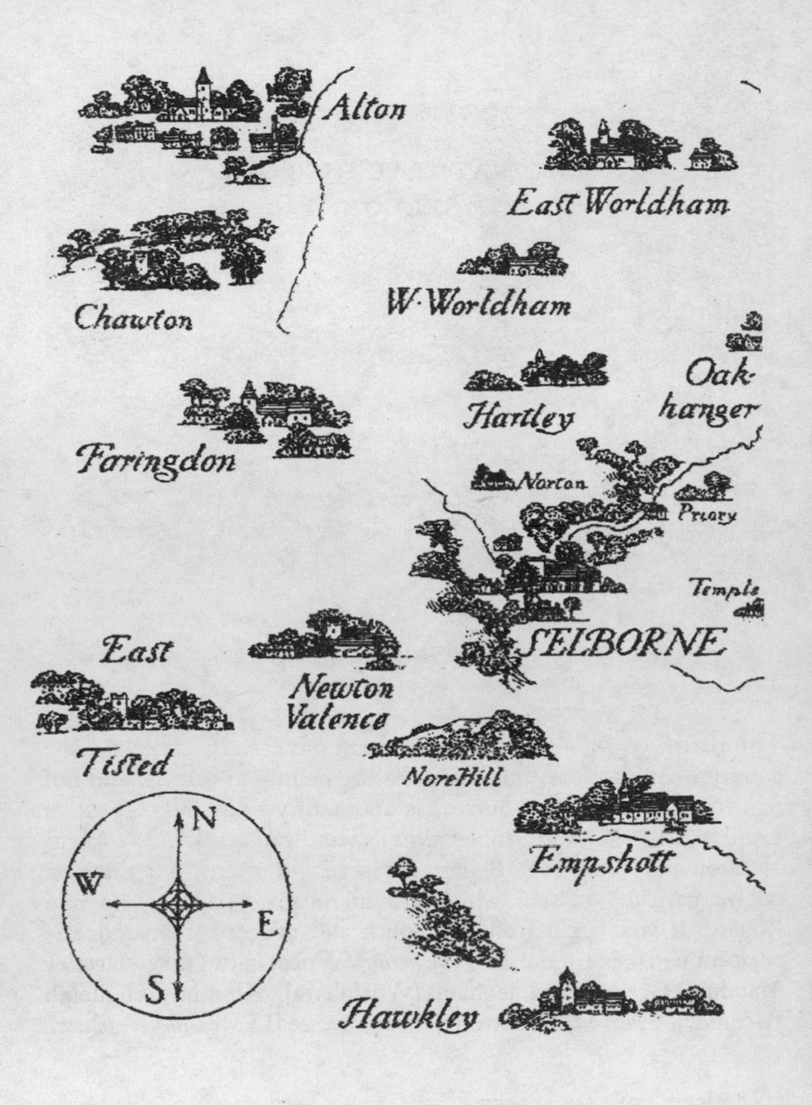
Alton
East Worldham
Chawton
W·Worldham
Oak-
hanger
Hartley
Faringdon
Norton
Priory
Temple
SELBORNE
East
Tisted
Newton
Valence
Nore Hill
Empshott
Hawkley
N
W
E
S

Selborne Church

Letter 1 to Thomas Pennant, Esq.[1]

The parish of Selborne lies in the extreme eastern corner of the county of Hampshire, bordering on the county of Sussex, and not far from the county of Surrey; is about fifty miles south-west of London, in latitude fifty-one, and near midway between the towns of Alton and Petersfield. Being very large and extensive it abuts on twelve parishes, two of which are in Sussex, viz., Trotton and Rogate. If you begin from the south and proceed westward, the adjacent parishes are Emshot, Newton Valence, Faringdon, Harteley Mauduit, Great Ward le ham [Worldham], Kingsley, Hedleigh [Headley], Bramshot, Trotton, Rogate, Lysse [Liss], and Greatham.

1 This letter, with some of those that follow it, seems never to have been written as such to the person to whom it was nominally addressed. It was probably added to the series by White when publication was finally decided upon, as explained in the Introduction. White evidently disliked the imputation of deliberate authorship, and modestly covered his prefatory matter with the cloak of correspondence. ED.

The soils of this district are almost as various and diversified as the views and aspects. The high part of the south-west consists of a vast hill of chalk, rising three hundred feet above the village, and is divided into a sheep-down, the high wood and a long hanging wood, called the Hanger. The covert of this eminence is altogether beech, the most lovely of all forest trees, whether we consider its smooth rind or bark, its glossy foliage, or graceful pendulous boughs. The down, or sheep-walk, is a pleasing park-like spot, of about one mile by half that space, jutting out on the verge of the hill-country, where it begins to break down into the plains, and commanding a very engaging view, being an assemblage of hill, dale, wood-lands, heath, and water. The prospect is bounded to the south-east and east by the vast range of mountains called the Sussex Downs, by Guild-down near Guildford, and by the Downs round Dorking, and Ryegate [Reigate] in Surrey, to the north-east, which altogether, with the country beyond Alton and Farnham, form a noble and extensive outline.

At the foot of this hill, one stage or step from the uplands, lies the village, which consists of one single straggling street, three-quarters of a mile in length, in a sheltered vale, and running parallel with the Hanger. The houses are divided from the hill by a vein of stiff clay (good wheat-land), yet stand on a rock of white stone;[2] little in appearance removed from chalk; but seems so far from being calcareous, that it endures extreme heat. Yet that the freestone still preserves somewhat that is analogous to chalk, is plain from the beeches which descend as low as those rocks extend, and no further, and thrive as well on them, where the ground is steep, as on the chalks.

The cartway of the village divides, in a remarkable manner, two very incongruous soils. To the south-west is a rank clay;[3] that requires the labour of years to render it mellow; while the gardens to the north-east, and small enclosures behind, consist of a warm, forward, crumbling mould, called black malm,[4] which seems highly saturated with vegetable and animal manure; and these may perhaps have been the original site of the town; while the woods and coverts might extend down to the opposite bank.

2 When White wrote, geology was hardly even in its infancy: the stone to which he here alludes is now known as one of the Upper Greensand series. ED.

3 Now called the Gault. ED

4 So known locally to the present day.

Selborne from the Hanger

At each end of the village, which runs from south-east to northwest, arises a small rivulet: that at the north-west end frequently fails; but the other is a fine perennial spring, little influenced by drought or wet seasons, called Well-head.[5] This breaks out of some high grounds joining to Nore Hill, a noble chalk promontory, remarkable for sending forth two streams into two different seas. The one to the south becomes a branch of the Arun, running to Arundel, and so falling into the British Channel: the other, to the north, the Selborne stream, makes one branch of the Wey; and, meeting the Black-down stream at Hedleigh [Headley] and the Alton and Farnham stream at Tilford-bridge, swells into a considerable river, navigable at Godalming; from whence it passes to Guildford, and so into the Thames at Weybridge; and thus at the Nore into the German Ocean.[6]

Our wells, at an average, run to about sixty-three feet, and when sunk to that depth seldom fail; but produce a fine limpid water, soft to the taste, and much commended by those who drink the pure element, but which does not lather well with soap.[7]

To the north-west, north and east of the village, is a range of fair enclosures, consisting of what is called white malm,[8] a sort of rotten or rubble stone, which, when turned up to the frost and rain, moulders to pieces, and becomes manure to itself.[9]

Still on to the north-east, and a step lower, is a kind of white land,[10] neither chalk nor clay, neither fit for pasture nor for the plough, yet kindly for hops, which root deep in the freestone, and

5 This spring produced, September 10, 1781, after a severe hot summer, and a preceding dry spring and winter, nine gallons of water in a minute, which is 540 in an hour, and 12,960, or 216 hogsheads, in twenty-four hours, or one natural day. At this time many of the wells failed, and all the ponds in the vales were dry.

6 In all the editions I have seen, the first included, this sentence and the previous one are made unintelligible by placing a full stop at the word 'north' and omitting the commas at 'other' and 'stream.' I have restored the passage as the author obviously intended it to read. Here and in several other places, indeed, I have ventured to amend the text by correcting what I take to be evident printer's errors in the first edition.

7 The water is hard, being strongly impregnated with lime from the chalk.

8 Now known as Chloritic Marl: it contains abundant nodules of phosphates, which give it great fertility. ED.

9 This soil produces good wheat and clover.

10 Lower Greensand. ED

have their poles and wood for charcoal growing just at hand. The white soil produces the brightest hops.

As the parish still inclines down towards Wolmer Forest, at the juncture of the clays and sand the soil becomes a wet, sandy loam, remarkable for timber, and infamous for roads. The oaks of Temple and Blackmoor stand high in the estimation of purveyors, and have furnished much naval timber; while the trees on the freestone grow large, but are what workmen call shaky, and so brittle as often to fall to pieces in sawing. Beyond the sandy loam the soil becomes a hungry lean sand, till it mingles with the forest; and will produce little without the assistance of lime and turnips.

In the village

At Norton Farm

Letter 2 also to Thomas Pennant, Esq.[1]

In the court of Norton farm-house, a manor farm to the north-west of the village, on the white malm, stood within these twenty years a broadleaved elm,[2] or wych hazel, *ulmus folio latissimo scabro* of Ray, which, though it had lost a considerable leading bough in the great storm in the year 1703, equal to a moderate tree, yet, when felled, contained eight loads of timber; and, being too bulky for a carriage, was sawn off at seven feet above the butt, where it measured near eight feet in the diameter. This elm I mention to show to what a bulk planted elms may attain; as this tree must certainly have been such from its situation.

1 This letter may possibly be an extract from one or more real letters written to Pennant. ED.

2 A wych-elm, *Ulmus montana*. ED.

In the centre of the village, and near the church, is a square piece of ground surrounded by houses, and vulgarly called 'The Plestor'.[3] In the midst of this spot stood, in old times, a vast oak, with a short squat body, and huge horizontal arms extending almost to the extremity of the area. This venerable tree, surrounded with stone steps, and seats above them, was the delight of old and young, and a place of much resort in summer evenings; where the former sat in grave debate, while the latter frolicked and danced before them. Long might it have stood, had not the amazing tempest in 1703[4] overturned it at once, to the infinite regret of the inhabitants, and the vicar, who bestowed several pounds in setting it in its place again: but all his care could not avail; the tree sprouted for a time, then withered and died. This oak I mention to show to what a bulk planted oaks also may arrive: and planted this tree must certainly have been, as will appear from what will be said farther concerning this area, when we enter on the antiquities of Selborne.

On the Blackmoor estate there is a small wood called Losel's, of a few acres, that was lately furnished with a set of oaks of a peculiar growth and great value; they were tall and taper like firs, but standing near together had very small heads, only a little brush without any large limbs. About twenty years ago the bridge at the Toy, near Hampton Court, being much decayed, some trees were wanted for the repairs that were fifty feet long without bough, and would measure twelve inches diameter at the little end. Twenty such trees did a purveyor find in this little wood, with this advantage, that many of them answered the description at sixty feet. These trees were sold for twenty pounds apiece.

In the centre of this grove there stood an oak, which, though shapely and tall on the whole, bulged out into a large excrescence about the middle of the stem. On this a pair of ravens had fixed their residence for such a series of years, that the oak was distinguished by the title of the Raven Tree. Many were the attempts of the neighbouring youths to get at this eyry: the difficulty whetted their inclinations, and each was ambitious of surmounting the

3 That is to say, the play-stow, or playing-place. ED

4 The great storm of 1703, the only one in Britain which (in historical times) has ever equalled the violence of a tropical hurricane, produced so deep an impression upon the people of the period that it was familiarly spoken of as 'the storm' throughout the whole of the eighteenth century. White, who was not born till seventeen years later, speaks of it as a well-known occurrence, both here and elsewhere. Macaulay gives a graphic description of this famous tempest in his essay on Addison. ED

The Plestor

arduous task. But, when they arrived at the swelling, it jutted out so in their way, and was so far beyond their grasp, that the most daring lads were awed, and acknowledged the undertaking to be too hazardous: so the ravens built on, nest upon nest, in perfect security, till the fatal day arrived in which the wood was to be levelled. It was in the month of February, when these birds usually sit. The saw was applied to the butt, – the wedges were inserted into the opening, – the woods echoed to the heavy blow of the beetle or mall or mallet, – the tree nodded to its fall; but still the dam sat on. At last, when it gave way, the bird was flung from her nest; and, though her parental affection deserved a better fate, was whipped down by the twigs, which brought her dead to the ground.

Well Head

Letter 3 to Thomas Pennant, Esq.[1]

The fossil-shells of this district, and sorts of stone, such as have fallen within my observation, must not be passed over in silence. And first I must mention, as a great curiosity, a specimen that was ploughed up in the chalky fields, near the side of the Down, and given to me for the singularity of its appearance, which, to an incurious eye, seems like a petrified fish of about four inches long, the cardo passing for an head and mouth. It is in reality a bivalve of the Linnaean genus of Mytilus, and the species of Crista Galli;[2] called by Lister, *Rastellum*; by Rumphius, *Ostreum plicatum minus*; by D'Argenville, *Auris Porci, s. Crista Galli*; and by those who make collections, Cock's Comb. Though I applied to several such in London, I never could meet with an entire specimen; nor could I

1 This letter on the fossils of Selborne is clearly a later insertion, and is a sufficiently perfunctory performance. ED.

2 White was mistaken in referring this fossil, of which he gives an illustration in the first edition, to the *Mytilus crista-galli* of Linnaeus. It is in reality *Ostraea carinata*, a characteristic mollusk of the Greensand. ED.

ever find in books any engraving from a perfect one. In the superb museum at Leicester House permission was given to me to examine for this article; and, though I was disappointed as to the fossil, I was highly gratified with the sight of several of the shells themselves in high preservation. This bivalve is only known to inhabit the Indian ocean, where it fixes itself to a zoophyte, known by the name Gorgonia. The curious foldings of the suture the one into the other, the alternate flutings or grooves, and the curved form of my specimen being much easier expressed by the pencil than by words, I have caused it to be drawn and engraved.

Ostræa carinata

Cornua Ammonis are very common about this village. As we were cutting an inclining path[3] up the Hanger, the labourers found them frequently on that steep, just under the soil, in the chalk, and of a considerable size. In the lane above Well-head, in the way to Emshot, they abound in the bank in a darkish sort of marl; and are usually very small and soft: but in Clay's Pond, a little further on, at the end of the pit, where the soil is dug out for manure, I have occasionally observed them of large dimensions, perhaps fourteen or sixteen inches in diameter. But as these did not consist of firm stone, but were formed of a kind of terra lapidosa, or hardened clay, as soon as they were exposed to the rains and frost they mouldered away. These seemed as if they were a very recent production. In the chalk-pit, at the north-west end of the Hanger, large nautili are sometimes observed.

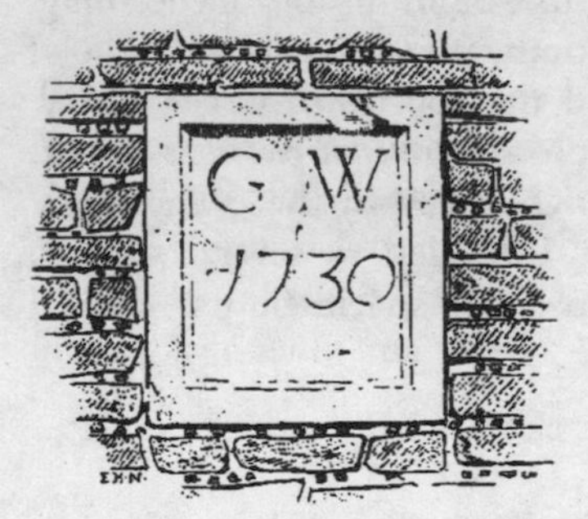

In the very thickest strata of our freestone, and at considerable depths, well-diggers often find large scallops or pectines, having both shells deeply striated, and ridged and furrowed alternately. They are highly impregnated with, if not wholly composed of, the stone of the quarry.

3 Doubtless the 'Bostal', constructed in 1780. E.H.N.

Letter 4 also to Thomas Pennant, Esq.[1]

As in a former letter the freestone of this place has been only mentioned incidentally, I shall here become more particular.

This stone is in great request for hearth-stones, and the beds of ovens: and in lining of lime-kilns it turns to good account; for the workmen use sandy loam instead of mortar; the sand of which fluxes,[2] and runs by the intense heat, and so cases over the whole face of the kiln with a strong vitrified coat like glass, that it is well preserved from injuries of weather, and endures thirty or forty years. When chiselled smooth, it makes elegant fronts for houses, equal in colour and grain to the Bath stone; and superior in one respect, that, when seasoned, it does not scale. Decent chimney-pieces are worked from it of much closer and finer grain than

1 Obviously an added letter. ED.

2 There may probably be also in the chalk itself that is burnt for lime a proportion of sand: for few chalks are so pure as to have none.

Portland; and rooms are floored with it; but it proves rather too soft for this purpose. It is a freestone cutting in all directions; yet has something of a grain parallel with the horizon, and therefore should not be surbedded, but laid in the same position that it grows in the quarry.[3]

On the ground abroad this firestone will not succeed for pavements, because, probably some degree of saltness prevailing within it, the rain tears the slabs to pieces.[4] Though this stone is too hard to be acted on by vinegar, yet both the white part, and even the blue rag, ferments strongly in mineral acids. Though the white stone will not bear wet, yet in every quarry at intervals there are thin strata of blue rag, which resists rain and frost; and are excellent for pitching of stables, paths and courts, and for building of dry walls against banks, a valuable species of fencing much in use in this village, and for mending of roads. This rag is rugged and stubborn, and will not hew to a smooth face, but is very durable; yet, as these strata are shallow and lie deep, large quantities cannot be procured but at considerable expense. Among the blue rags turn up some blocks tinged with a stain of yellow or rust colour, which seem to be nearly as lasting as the blue; and every now and then balls of a friable substance, like rust of iron, called rust balls.

In Wolmer Forest I see but one sort of stone, called by the workmen sand, or forest-stone. This is generally of the colour of rusty iron, and might probably be worked as iron ore; is very hard and heavy, and of a firm, compact texture, and composed of a small roundish crystalline grit, cemented together by a brown, terrene, ferruginous matter; will not cut without difficulty, nor easily strike fire with steel. Being often found in broad flat pieces, it makes good pavement for paths about houses, never becoming slippery in frost or rain; is excellent for dry walls, and is sometimes used in buildings. In many parts of that waste it lies scattered on the surface of the ground; but is dug on Weaver's Down, a vast hill on the eastern verge of that forest, where the pits are shallow and the stratum thin. This stone is imperishable.

3 To *surbed* stone is to set it edgewise, contrary to the posture it had in the quarry, says Dr Plot, *Oxfordshire*, p. 77. But *surbedding* does not succeed in our dry walls; neither do we use it so in ovens, though he says it is best for Teynton stone.

4 'Firestone is full of salts, and has no sulphur: must be close-grained, and have no interstices. Nothing supports fire like salts; saltstone perishes exposed to wet and frost.' – PLOT, *History of Staffordshire*, p. 152.

A corner of Whites house

The Plestor in 1776

From a notion of rendering their work the more elegant, and giving it a finish, masons chip this stone into small fragments about the size of the head of a large nail, and then stick the pieces into the wet mortar along the joints of their freestone walls; this embellishment carries an odd appearance, and has occasioned strangers sometimes to ask us pleasantly, 'whether we fastened our walls together with tenpenny nails'.[5]

5 Walls of this sort still occur at Selborne: there are many close to the church. They are also common at Dorking and in other places on the Greensand area. For an example, see illustration on page 14. ED.

Hollow Lane near Norton

Letter 5 also to Thomas Pennant, Esq.[1]

Among the singularities of this place the two rocky hollow lanes, the one to Alton, and the other to the forest, deserve our attention. These roads, running through the malm lands, are, by the traffic of ages, and the fretting of water, worn down through the first stratum of our freestone, and partly through the second; so that they look more like water-courses than roads; and are bedded with naked rag for furlongs together. In many places they are reduced sixteen or eighteen feet beneath the level of the fields; and after floods, and in frosts, exhibit very grotesque and wild appearances, from the tangled roots that are twisted among the strata, and from the torrents rushing down their broken sides; and especially when those cascades are frozen into icicles, hanging in all the fanciful shapes of frostwork. These rugged gloomy scenes affright the ladies when

1 A made-up letter on the roads and human aspects of Selborne. ED.

North-east view of Selborne from the

Short Lythe in Gilbert White's time

they peep down into them from the paths above, and make timid horsemen shudder while they ride along them; but delight the naturalist with their various botany, and particularly with their curious filices, with which they abound.

The manor of Selborne, was it strictly looked after, with all its kindly aspects, and all its sloping coverts, would swarm with game; even now hares, partridges, and pheasants abound; and in old days woodcocks were as plentiful. There are few quails, because they more affect open fields than enclosures; after harvest some few landrails are seen.

The parish of Selborne, by taking in so much of the forest, is a vast district. Those who tread the bounds are employed part of three days in the business, and are of opinion that the outline, in all its curves and indentings, does not comprise less than thirty miles.

The village stands in a sheltered spot, secured by the Hanger from the strong westerly winds. The air is soft, but rather moist from the effluvia of so many trees; yet perfectly healthy and free from agues.

The quantity of rain that falls on it is very considerable, as may be supposed in so woody and mountainous a district. As my experience in measuring the water is but of short date, I am not qualified to give the mean quantity.[2] I only know that:

		INCHES
from	May 1, 1779, to the end of the year there fell	28.37!
	January 1, 1780, to January 1, 1781	27.32
	January 1, 1781, to January 1, 1782	30.71
	January 1, 1782, to January 1, 1783	50.26!
	January 1, 1783, to January 1, 1784	33.71
	January 1, 1784, to January 1, 1785	33.80
	January 1, 1785, to January 1, 1786	31.55
	January 1, 1786, to January 1, 1787	39.57

2 A very intelligent gentleman assures me (and he speaks from upwards of forty years' experience), that the mean rain of any place cannot be ascertained till a person has measured it for a very long period. 'If I had only measured the rain,' says he, 'for the four first years, from 1740 to 1743, I should have said the mean rain at Lyndon was 16 ½ inches for the year; if from 1740 to 1750, 18 ½ inches. The mean rain before 1763 was 20¼ inches, from 1763 and since 25 ½ inches, from 1770 to 1780, 26 inches. If only 1773, 1774, and 1775, had been measured, Lyndon mean rain would have been called 32 inches.'

The village of Selborne, and large hamlet of Oakhanger, with the single farms, and many scattered houses along the verge of the forest, contain upwards of six hundred and seventy inhabitants.[3]

3 *A State of the Parish of Selborne, taken October 4, 1783*

The number of tenements or families,	136.
The number of inhabitants in the street is	313
In the rest of the parish	363
Total	676

(*near five inhabitants to each tenement*.)

In the time of the Revd Gilbert White, vicar, who died in 1727-8, the number of inhabitants was computed at about 500.

Average number of Baptisms annually over 60 years

Period	Male	Female	Total
From 1720 to 1729 inclusive	6.9	6.0	12.9
1730 to 1739 inclusive	6.9	8.4	15.3
1740 to 1749 inclusive	9.2	6.6	15.8
1750 to 1759 inclusive	7.6	8.1	15.7
1760 to 1769 inclusive	9.1	8.9	18.0
1770 to 1779 inclusive	10.5	9.8	20.3
Total number of baptisms from 1720 to 1779, inclusive – 60 years	515	465	980

Average number of Burials annually over 60 years

Period	Male	Female	Total
From 1720 to 1729 inclusive	4.8	5.1	9.9
1730 to 1739 inclusive	4.8	5.8	10.6
1740 to 1749 inclusive	4.6	3.8	8.4
1750 to 1759 inclusive	4.9	5.1	10.0
1760 to 1769 inclusive	6.9	6.5	13.4
1770 to 1779 inclusive	5.5	6.2	11.7
Total number of burials from 1720 to 1779, inclusive – 60 years	315	325	640

Baptisms exceed burials by more than one-third.

Baptisms of Males exceed Females by one-tenth, or one in ten.

Burials of Females exceed Males by one in thirty.

It appears that a child, born and bred in this parish, has an equal chance to live above forty years.

Twins thirteen times, many of whom dying young have lessened the chance for life.

Chances for life in men and women appear to be equal.

We abound with poor; many of whom are sober and industrious, and live comfortably in good stone or brick cottages, which are glazed, and have chambers; above stairs: mud buildings we have none. Besides the employment from husbandry, the men work in hop-gardens, of which we have many; and fell and bark timber. In the spring and summer the women weed the corn; and enjoy a second harvest in September by hop-picking. Formerly, in the dead months they availed themselves greatly by spinning wool, for

---

The Baptisms, Burials and Marriages from January 2, 1761 to December 25, 1780 in the Parish of Selborne

	Baptisms			Burials			Marriages
	Male	Female	Total	Male	Female	Total	
1761	8	10	18	2	4	6	3
1762	7	8	15	10	14	24	6
1763	8	10	18	3	4	7	5
1764	11	9	20	10	8	18	6
1765	12	6	18	9	7	16	6
1766	9	13	22	10	6	16	4
1767	14	5	19	6	5	11	2
1768	7	6	13	2	5	7	6
1769	9	14	23	6	5	11	2
1770	10	13	23	4	7	11	3
1771	10	6	16	3	4	7	4
1772	11	10	21	6	10	16	3
1773	8	5	13	7	5	12	3
1774	6	13	19	2	8	10	1
1775	20	7	27	13	8	21	6
1776	11	10	21	4	6	10	6
1777	8	13	21	7	3	10	4
1778	7	13	20	3	4	7	5
1779	14	8	22	5	6	11	5
1780	8	9	17	11	4	15	3
	198	188	386	123	123	246	83

During this period of twenty years the births of males exceeded those off females by 10

And the births exceeded the deaths by 140

The burials of each sex were equal.

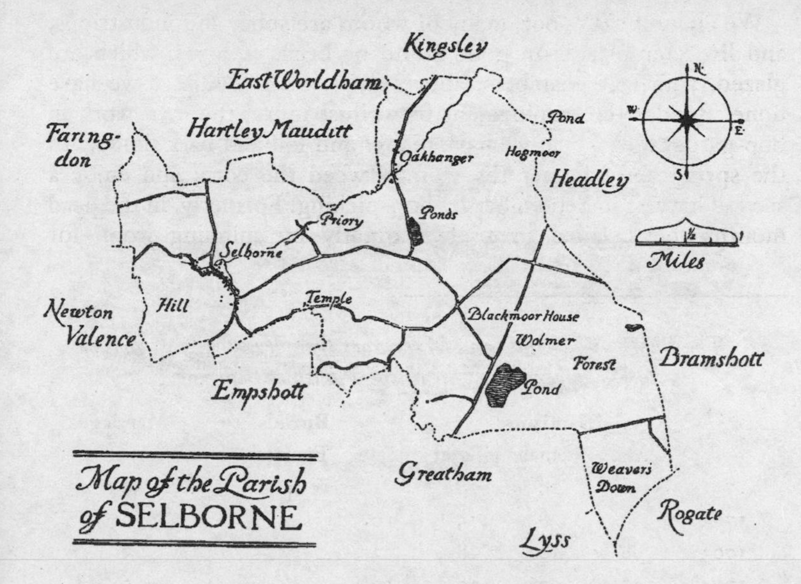

making of barragons, a genteel corded stuff, much in vogue at that time for summer wear; and chiefly manufactured at Alton, a neighbouring town, by some of the people called Quakers: but from circumstances this trade is at an end.[4] The inhabitants enjoy a good share of health and longevity; and the parish swarms with children.

4 Since the passage above was written, I am happy in being able to say that the spinning employment is a little revived, to the no small comfort of the industrious housewife.

In Wolmer Forest

Letter 6 also to Thomas Pennant, Esq.

Should I omit to describe with some exactness the forest of Wolmer, of which three-fifths perhaps lie in this parish, my account of Selborne would be very imperfect, as it is a district abounding with many curious productions, both animal and vegetable; and has often afforded me much entertainment both as a sportsman and as a naturalist.

The royal forest of Wolmer is a tract of land of about seven miles in length, by two and a half in breadth, running nearly from north to south, and is abutted on, to begin to the south, and so to proceed eastward, by the parishes of Greatham, Lysse [Liss], Rogate, and Trotton, in the county of Sussex; by Bramshot, Hedleigh [Headley], and Kingsley. This royalty consists entirely of sand covered with heath and fern; but it is somewhat diversified with hills and dales, without having one standing tree in the whole extent. In the bottoms, where the waters stagnate, are many bogs, which formerly abounded with subterraneous trees; though Dr Plot says positively,[1] that 'there never were any fallen trees hidden in the mosses of the

1 See his *History of Staffordshire*.

southern counties.' But he was mistaken: for I myself have seen cottages on the verge of this wild district, whose timbers consisted of a black hard wood, looking like oak, which the owners assured me they procured from the bogs by probing the soil with spits, or some such instruments: but the peat is so much cut out, and the moors have been so well examined, that none has been found of late.[2] Besides the oak, I have also been shown pieces of fossil wood of a paler colour, and softer nature, which the inhabitants called fir: but, upon a nice examination, and trial by fire, I could discover nothing resinous in them; and therefore rather suppose that they were parts of a willow or alder, or some such aquatic tree.[3]

This lonely domain is a very agreeable haunt for many sorts of wild fowls, which not only frequent it in the winter, but breed there in the summer; such as lapwings, snipes, wild-ducks, and, as I have discovered within these few years, teals. Partridges in vast plenty are

2 Old people have assured me, that on a winter's morning they have discovered these trees, in the bogs, by the hoar frost, which lay longer over the space where they were concealed than in the surrounding morass. Nor does this seem to be a fanciful notion, but consistent with true philosophy. Dr Hales saith, 'That the warmth of the earth, at some depth underground, has an influence in promoting a thaw, as well as the change of the weather from a freezing to a thawing state, is manifest, from this observation, viz., November 29, 1731, a little snow having fallen in the night, it was, by eleven the next morning, mostly melted away on the surface of the earth, except in several places in Bushy Park, where there were drains dug and covered with earth, on which the snow continued to lie, whether those drains were full of water or dry; as also where elm-pipes lay underground: a plain proof this, that those drains intercepted the warmth of the earth from ascending from greater depths below them; for the snow lay where the drain had more than four feet depth of earth over it. It continued also to lie on thatch, tiles, and the tops of walls.' – See Hales's *Haemastatics*, p. 360. *Query*, Might not such observations be reduced to domestic use, by promoting the discovery of old obliterated drains and wells about houses; and in Roman stations and camps lead to the finding of pavements, baths, and graves, and other hidden relics of curious antiquity?

3 Many errors still occur even among 'educated' people from a misconception of the meaning of the word *forest*. In early times, it did not necessarily or even usually imply the presence of trees. A forest is not a wooded district but one reserved for hunting and sport – what we now call a moor or heath. Wolmer Forest was never wooded at ancient dates, though a small part of it has lately been planted with Scotch firs. Legends of trees having once existed on bare tracts have often grown up through a misapprehension of the meaning of the word *forest*. Deer-forests cannot, of course, be thickly wooded: the word is used in this case in its original and proper meaning. ED.

bred in good seasons on the verge of this forest, into which they love to make excursions: and in particular, in the dry summer of 1740 and 1741, and some years after, they swarmed to such a degree that parties of unreasonable sportsmen killed twenty and sometimes thirty brace in a day.

Blackcock

But there was a nobler species of game in this forest, now extinct, which I have heard old people say abounded much before shooting flying became so common, and that was the heath-cock, black game, or grouse. When I was a little boy I recollect one coming now and then to my father's table. The last pack remembered was killed about thirty-five years ago; and within these ten years one solitary greyhen was sprung by some beagles in beating for a hare. The sportsmen cried out, 'A hen pheasant'; but a gentleman present, who had often seen grouse in the north of England, assured me that it was a greyhen.

Nor does the loss of our black game prove the only gap in the Fauna Selborniensis; for another beautiful link in the chain of beings is wanting, I mean the red deer, which toward the beginning of this century amounted to about five hundred head, and made a stately appearance. There is an old keeper, now alive, named Adams, whose great-grandfather (mentioned in a perambulation taken in 1635), grandfather, father, and self, enjoyed the head keepership of Wolmer Forest in succession for more than an hundred years. This person assures me, that his father has often told him, that Queen Anne, as she was journeying on the Portsmouth road, did not think the forest of Wolmer beneath her royal regard. For she came out of the great road[1] at Lippock [Liphook], which is just by, and reposing herself on a bank smoothed for that purpose, lying about half a mile to the east of Wolmer Pond, and still called Queen's Bank, saw with great complacency and satisfaction the whole herd of red deer brought by the keepers along the vale before her, consisting then of about five hundred head. A sight this, worthy the attention of the greatest sovereign! But he farther adds that, by means of the Waltham blacks,[2] or, to use his own expression, as

1 The Portsmouth Road. ED.

2 A body of local deer-stealers or poachers, for details as to whom. see the next letter. ED

Wolmer Pond

soon as they began blacking, they were reduced to about fifty head, and so continued decreasing till the time of the late Duke of Cumberland. It is now more than thirty years ago that his Highness sent down an huntsman, and six yeoman-prickers, in scarlet jackets laced with gold, attended by the stag-hounds; ordering them to take every deer in this forest alive, and to convey them in carts to Windsor. In the course of the summer they caught every stag, some of which showed extraordinary diversion: but in the following winter, when the hinds were also carried off, such fine chases were exhibited as served the country people for matter of talk and wonder for years afterwards. I saw myself one of the yeoman-prickers single out a stag from the herd, and must confess that it was the most curious feat of activity I ever beheld, superior to anything in Mr Astley's riding-school. The exertions made by the horse and deer much exceeded all my expectations; though the former greatly excelled the latter in speed. When the devoted deer was separated from his companions, they gave him, by their watches, law, as they called it, for twenty minutes; when, sounding their horns, the stopdogs were permitted to pursue, and a most gallant scene ensued.

View of Selborne

Letter 7 also to Thomas Pennant, Esq.[1]

Though large herds of deer do much harm to the neighbourhood, yet the injury to the morals of the people is of more moment than the loss of their crops. The temptation is irresistible; for most men are sportsmen by constitution: and there is such an inherent spirit for hunting in human nature, as scarce any inhibitions can restrain. Hence, towards the beginning of this century all this country was wild about deer-stealing. Unless he was a hunter, as they affected to call themselves, no young person was allowed to be possessed of manhood or gallantry. The Waltham blacks at length committed such enormities, that government was forced to interfere with that severe and sanguinary act called the 'Black Act',[2] which now comprehends more felonies than any law that ever was framed before. And, therefore, a late Bishop of Winchester, when urged to re-stock Waltham Chase,[3] refused, from a motive worthy of a prelate, replying 'that it had done mischief enough already.'

1 Also an added letter, suggested by the need for explanation of the last. ED.

2 Statute 9 Geo. I. cap. 22.

3 This chase remains unstocked to this day; the bishop was Dr Hoadly.

from the North

Our old race of deer-stealers are hardly extinct yet: it was but a little while ago that, over their ale, they used to recount the exploits of their youth; such as watching the pregnant hind to her lair, and, when the calf was dropped, paring its feet with a penknife to the quick to prevent its escape, till it was large and fat enough to be killed; the shooting at one of their neighbours with a bullet in a turnip-field by moonshine, mistaking him for a deer; and the losing a dog in the following extraordinary manner: – Some fellows, suspecting that a calf new-fallen was deposited in a certain spot of thick fern, went, with a lurcher, to surprise it; when the parent-hind rushed out of the brake, and, taking a vast spring with all her feet close together, pitched upon the neck of the dog, and broke it short in two.

Another temptation to idleness and sporting was a number of rabbits, which possessed all the hillocks and dry places: but these being inconvenient to the huntsmen, on account of their burrows, when they came to take away the deer, they permitted the country people to destroy them all.

Such forests and wastes, when their allurements to irregularities are removed, are of considerable service to the neighbourhoods that verge upon them, by furnishing them with peat and turf for their firing; with fuel for the burning their lime; and with ashes for their grasses; and by maintaining their geese and their stock of young cattle at little or no expense.

The manor-farm of the parish of Greatham has an admitted

claim, I see (by an old record taken from the Tower of London), of turning all live stock on the forest, at proper seasons, 'bidentibus exceptis'.[4] The reason, I presume, why sheep[5] are excluded, is, because, being such close grazers, they would pick out all the finest grasses, and hinder the deer from thriving.

Though (by statute 4 and 5 W. and Mary, c. 23) 'to burn on any waste, between Candlemas and Midsummer, any grig, ling, heath and furze, goss or fern, is punishable with whipping and confinement in the house of correction'; yet, in this forest, about March or April, according to the dryness of the season, such vast heath-fires are lighted up, that they often get to a masterless head, and, catching the hedges, have sometimes been communicated to the underwoods, woods, and coppices, where great damage has ensued. The plea for these burnings is, that, when the old coat of heath, &c., is consumed, young will sprout up, and afford much tender brouze for cattle; but, where there is large old furze, the fire, following the roots, consumes the very ground; so that for hundreds of acres nothing is to be seen but smother and desolation, the whole circuit round looking like the cinders of a volcano; and, the soil being quite exhausted, no traces of vegetation are to be found for years. These conflagrations, as they take place usually with a north-east or east wind, much annoy this village with their smoke, and often alarm the country; and, once in particular, I remember that a gentleman, who lives beyond Andover, coming to my house, when he got on the downs between that town and Winchester, at twenty-five miles distance, was surprised much with smoke and a hot smell of fire; and concluded that Alresford was in flames; but, when he came to that town, he then had apprehensions for the next village, and so on to the end of his journey.

On two of the most conspicuous eminences of this forest stand two arbours or bowers, made of the boughs of oaks; the one called Waldon Lodge, the other Brimstone Lodge: these the keepers renew annually on the feast of St. Barnabas, taking the old materials for a perquisite. The farm called Blackmoor, in this parish, is obliged to find the posts and brush-wood for the former; while the farms at Greatham, in rotation, furnish for the latter; and are all enjoined to cut and deliver the materials at the spot. This custom I mention, because I look upon it to be of very remote antiquity.

4 For this privilege the owners of that estate used to pay to the king annually seven bushels of oats.

5 In the Holt, where a full stock of fallow-deer has been kept up till lately, no sheep are admitted to this day.

Oakhanger Pond

Letter 8 also to Thomas Pennant, Esq.

On the verge of the forest, as it is now circumscribed, are three considerable lakes, two in Oakhanger, of which I have nothing particular to say; and one called Bin's, or Bean's Pond, which is worthy the attention of a naturalist or a sportsman. For being crowded at the upper end with willows, and with the carex cespitosa[1] it affords such a safe and pleasing shelter to wild ducks, teals,snipes, &c., that they breed there. In the winter this covert is also frequented by foxes, and sometimes by pheasants; and the bogs produce many curious plants.

By a perambulation of Wolmer Forest and the Holt, made in 1735 and the eleventh year of Charles the First (which now lies before me), it appears that the limits of the former are much circumscribed.

1 I mean that sort which, rising into tall hassocks, is called by the foresters torrets; a corruption, I suppose, of turrets.

For, to say nothing of the farther side, with which I am not so well acquainted, the bounds on this side, in old times, came into Binswood; and extended to the ditch of Ward le Ham Park, in which stands the curious mount called King John's Hill and Lodge Hill; and to the verge of Hartley Mauduit, called Mauduit Hatch; comprehending also Short Heath, Oakhanger, and Oakwoods; a large district, now private property, though once belonging to the royal domain.[2]

It is remarkable that the term purlieu is never once mentioned in this long roll of parchment. It contains, besides the perambulation, a rough estimate of the value of the timbers, which were considerable, growing at that time in the district of the Holt; and enumerates the officers, superior and inferior, of those joint forests, for the time being, and their ostensible fees and perquisites. In those days as at present, there were hardly any trees in Wolmer Forest.

Within the present limits of the forest are three considerable lakes, Hogmer, Cranmer, and Wolmer; all of which are stocked with carp, tench, eels, and perch: but the fish do not thrive well because the water is hungry, and the bottoms are a naked sand.

A circumstance respecting these ponds, though by no means peculiar to them, I cannot pass over in silence; and that is, that instinct by which in summer all the kine, whether oxen, cows calves, or heifers, retire constantly to the water during the hotter hours; where, being more exempt from flies, and inhaling the coolness of that element, some belly deep, and some only to mid-leg, they ruminate and solace themselves from about ten in the morning till four in the afternoon, and then return to their feeding. During this great proportion of the day they drop much dung, in which insects nestle; and so supply food for the fish, which would be poorly subsisted but from this contingency. Thus Nature, who is a great economist, converts the recreation of one animal to the support of another! Thomson, who was a nice observer of natural occurrences, did not let this pleasing circumstance escape him. He says, in his *Summer*,

> A various group the herds and flocks compose;
> . . . on the grassy bank
> Some ruminating lie; while others stand
> Half in the flood, and, often bending, sip
> The circling surface.

2 In the beginning of the summer 1787, the royal forests of Wolmer and Holt were measured by persons sent down by government.

Wolmer Pond, so called, I suppose, for eminence sake;[3] is a vast lake for this part of the world, containing, in its whole circumference, 2646 yards, or very near a mile and a half. The length of the north-west and opposite side is about 704 yards, and the breadth of the south-west end about 456 yards. This measurement, which I caused to be made with good exactness, gives an area of about sixty-six acres, exclusive of a large irregular arm at the north-east corner, which we did not take into the reckoning.

On the face of this expanse of waters, and perfectly secure from fowlers, lie all day long, in the winter season, vast flocks of ducks, teals, and widgeons, of various denominations; where they preen and solace, and rest themselves, till towards sunset, when they issue forth in little parties (for in their natural state they are all birds of the night) to feed in the brooks and meadows; returning again with the dawn of the morning. Had this lake an arm or two more, and were it planted round with thick covert (for now it is perfectly naked), it might make a valuable decoy.

Yet neither its extent, nor the clearness of its water, nor the resort of various and curious fowls, nor its picturesque groups of cattle, can render this meer so remarkable as the great quantity of coins that were found in its bed about forty years ago. But, as such discoveries more properly belong to the antiquities of this place, I shall suppress all particulars for the present, till I enter professedly on my series of letters respecting the more remote history of this village and district.[4]

3 White is mistaken, I need hardly say, in supposing the pond to be called after the forest: it is really the forest which is called after the pond. The wild tract between Petersfield, Haslemere, and Selborne contained three meres, Hogmere, Cranmere, and Wolmere, or Hogmer, Cranmer, and Wolmer. From the largest of these three, Wolmer, the forest took its usual name. Wolmer Pond was once much larger than in White's time, and has now been still more extensively drained, till it is quite insignificant. ED.

4 These letters, though included in the first edition, form a separate work, under the title of *The Antiquities of Selborne*. ED.

In Alice Holt Forest

Letter 9 also to Thomas Pennant, Esq.

By way of supplement, I shall trouble you once more on this subject, to inform you that Wolmer, with her sister forest, Ayles Holt, alias Alice Holt,[1] as it is called in old records, is held by grant from the crown for a term of years.

The grantees that the author remembers are Brigadier-General Emanuel Scroope Howe, and his lady, Ruperta, who was a natural daughter of Prince Rupert by Margaret Hughes; a Mr Mordaunt, of the Peterborough family, who married a dowager Lady Pembroke; Henry Bilson Legge and lady; and now Lord Stawell, their son.

The lady of General Howe lived to an advanced age, long surviving her husband; and, at her death, left behind her many

1 In *Rot. Inquisit de statu forest. in Scaccar*. 36 Edward III, it is called Aisholt. In the same, '*Tit. Woolmer and Aisholt Hantisc. Dominus Rex habet unam capellam in haia sua de Kingesle*.' '*Haia, sepes, sepimentum, parcus; a Gall. haie and haye*.' SPELMAN'S *Glossary*.

curious pieces of mechanism of her father's constructing, who was a distinguished mechanic and artist[2] as well as warrior; and among the rest, a very complicated clock, lately in possession of Mr Elmer, the celebrated game painter at Farnham, in the county of Surrey.

Though these two forests are only parted by a narrow range of enclosures, yet no two soils can be more different; for the Holt consists of a strong loam, of a miry nature, carrying a good turf, and abounding with oaks that grow to be large timber; while Wolmer is nothing but a hungry, sandy, barren waste.

The former being all in the parish of Binsted, is about two miles in extent from north to south, and near as much from east to west; and contains within it many woodlands and lawns, and the great lodge where the grantees reside, and a smaller lodge called Goose Green; and is abutted on by the parishes of Kingsley, Frinsham, Farnham, and Bentley; all of which have right of common.

One thing is remarkable, that though the Holt has been of old well stocked with fallow-deer, unrestrained by any pales or fences more than a common hedge, yet they were never seen within the limits of Wolmer; nor were the red deer of Wolmer ever known to haunt the thickets or glades of the Holt.

At present the deer of the Holt are much thinned and reduced by the night hunters, who perpetually harass them in spite of the efforts of numerous keepers, and the severe penalties that have been put in force against them as often as they have been detected, and rendered liable to the lash of the law. Neither fines nor imprisonments can deter them; so impossible is it to extinguish the spirit of sporting which seems to be inherent in human nature.

General Howe turned out some German wild boars and sows in his forests, to the great terror of the neighbourhood, and, at one time, a wild bull or buffalo; but the country rose upon them and destroyed them.

2 This prince was the inventor of mezzotinto.

A very large fall of timber, consisting of about one thousand oaks, has been cut this spring (viz., 1784, in the Holt forest: one fifth of which, it is said, belongs to the grantee, Lord Stawell. He lays claim also to the lop and top; but the poor of the parishes of Binsted and Frinsham [Frensham], Bentley and Kingsley, assert that it belongs to them, and assembling in a riotous manner, have actually taken it all away. One man, who keeps a team, has carried home for his share forty sacks of wood. Forty-five of these people his lordship has served with actions. These trees, which were very sound and in high perfection, were winter-cut, viz., in February and March, before the bark would run. In old times the Holt was estimated to be eighteen miles, computed measure from water-carriage, viz., from the town of Chertsey, on the Thames; but now it is not half that distance, since the Wey is made navigable up to the town of Godalming in the county of Surrey.

Gilbert White's house

Letter 10 also to Thomas Pennant, Esq.[1]

August 4th, 1767

It has been my misfortune never to have had any neighbours whose studies have led them towards the pursuit of natural knowledge; so that, for want of a companion to quicken my industry and sharpen my attention, I have made but slender progress in a kind of information to which I have been attached from my childhood.

As to swallows (*hirundines rusticae*) being found in a torpid state during the winter in the Isle of Wight or any part of this country, I never heard any such account worth attending to. But a clergyman, of an inquisitive turn, assures me, that when he was a great boy,

1 With this letter, the first bearing a date, we begin the real series of White's interesting and valuable correspondence with Pennant. See Introduction. The style of the true letters is far superior to that of the artificial additions. ED.

some workmen, in pulling down the battlements of a church tower early in the spring, found two or three swifts (*hirundines apodes*) among the rubbish, which were at first appearance dead, but on being carried towards the fire revived. He told me, that out of his great care to preserve them, he put them in a paper bag, and hung them by the kitchen fire, where they were suffocated[2]

Another intelligent person has informed me, that while he was a schoolboy at Brighthelmstone [Brighton], in Sussex, a great fragment of the chalk cliff fell down one stormy winter on the beach, and that many people found swallows among the rubbish; but on my questioning him whether he saw any of those birds himself, to my no small disappointment, he answered me in the negative; but that others assured him they did.

Young broods of swallows began to appear this year on July the 11th, and young martins (*hirundines urbicae*) were then fledged in their nests. Both species will breed again once. For I see by my fauna of last year, that young broods came forth so late as September the 18th. Are not these late hatchings more in favour of hiding than migration? Nay, some young martins remained in their nests last year so late as September the 29th; and yet they totally disappeared with us by the 5th of October.

How strange it is that the swift, which seems to live exactly the same life with the swallow and house-martin, should leave us before the middle of August invariably! while the latter stay often till the middle of October; and once I saw numbers of house-martins on the 7th of November. The martins and red-wing fieldfares were flying in sight together, an uncommon assemblage of summer and winter birds!

2 This question whether swallows and their like were to be found hibernating in England seems to be the one about which Pennant first put himself in communication with the Selborne naturalist. It was commonly believed at the time that swallows were often found torpid in England, and even that they passed the winter under water in the mud of ponds. It is now known, of course, that such stories are quite untrue, and that swallows and swifts migrate southward in winter. The swift, again, is not related to the swallow, but is a Cypselus, belonging to an entirely different family. But White could never quite get over the belief in hibernation, a point to which he recurs again and again throughout these letters. ED.

A little yellow bird (it is either a species of the *alauda trivialis*, or rather perhaps of the *motacilla trochilus*) still continues to make a sibilous shivering noise in the tops of tall woods.[3] The stoparola of Ray (for which we have as yet no name in these parts) is called in your zoology the fly-catcher.[4] There is one circumstance characteristic of this bird which seems to have escaped observation, and that is, its takes its stand on the top of some stake or post, from whence it springs forth on its prey, catching a fly in the air, and hardly ever touching the ground, but returning still to the same stand for many times together.

Fly-catcher

I perceive there are more than one species of the *motacilla trochilus*. Mr Derham supposes, in Ray's *Philosophical Letters*, that he has discovered three. In these there is again an instance of some very common birds that have as yet no English name.

Mr Stillingfleet makes a question whether the black-cap (*motacilla atricapilla*) be a bird of passage or not: I think there is no doubt of it: for, in April, in the first fine weather, they come trooping, all at once, into these parts, but are never seen in the winter. They are delicate songsters.

Numbers of snipes breed every summer in some moory [marshy] ground on the verge of this parish. It is very amusing to see the cock bird on wing at that time, and to hear his piping and humming notes.

I have had no opportunity yet of procuring any of those mice[5] which I mentioned to you in town. The person that brought me the last says they are plenty in harvest, at which time I will take care to get more; and will endeavour to put the matter out of doubt whether it be a nondescript species or not.

I suspect much there may be two species of water-rats. Ray says, and Linnaeus after him, that the water-rat is web-footed behind. Now I have discovered a rat on the banks of our little stream that is not web-footed, and yet is an excellent swimmer and diver: it answers exactly to the *mus amphibius* of Linnaeus (see *Syst. Nat.*)

3 The yellow willow-wren, *Sylvia sibilatrix* (*Phylloscopus sibilatrix*). ED.
4 The spotted fly-catcher, *Muscicapa grisola*. ED.
5 Harvest-mice. ED.

which he says '*natat in fossis et urinatur*'. I should be glad to procure one '*plantis palmatis*'. Linnaeus seems to be in a puzzle about his *mus amphibius*, and to doubt whether it differs from his *mus terrestris*; which if it be, as he allows, the '*mus agrestis capite grandi brachyuros*', of Ray, is widely different from the water-rat, both in size, make, and manner of life.[6]

Water Vole

As to the *falco*, which I mentioned in town, I shall take the liberty to send it down to you into Wales; presuming on your candour that you will excuse me if it should appear as familiar to you as it is strange to me. Though mutilated '*qualem dices . . . antehac fuisse, tales cum sint reliquiae!*'

It haunted a marshy piece of ground in quest of wild-ducks and snipes; but, when it was shot, had just knocked down a rook, which it was tearing in pieces. I cannot make it answer to any of our English hawks; neither could I find any like it at the curious exhibition of stuffed birds in Spring Gardens. I found it nailed up at the end of a barn, which is the countryman's museum.

The parish I live in is a very abrupt, uneven country, full of hills and woods, and therefore full of birds.

6 We have only one so-called water-rat in Britain, better described as the water-vole, *Arvicola amphibius*; it is not web-footed. ED.

Hoopoe

Letter 11 also to Thomas Pennant, Esq.

Selborne, September 9th, 1767

It will not be without impatience that I shall wait for your thoughts with regard to the *falco*; as to its weight, breadth, &c., I wish I had set them down at the time; but, to the best of my remembrance, it weighed two pounds and eight ounces, and measured, from wing to wing, thirty-eight inches. Its cere and feet were yellow, and the circle of its eyelids a bright yellow. As it had been killed some days, and the eyes were sunk, I could make no good observation on the colour of the pupils and the irides.[1]

The most unusual birds I ever observed in these parts were a pair of hoopoes (*upupa*), which came several years ago in the summer, and frequented an ornamented piece of ground, which joins to my garden, for some weeks. They used to march about in a stately manner, feeding in the walks, many times in the day; and seemed disposed to breed in my outlet; but were frighted and persecuted by idle boys, who would never let them be at rest.

Three grossbeaks (*loxia coccothraustes*) appeared some years ago in

1 It was a peregrine falcon, *Falco peregrinus*. ED.

my fields, in the winter; one of which I shot. Since that, now and then, one is occasionally seen in the same dead season.

A crossbill (*loxia curvirostra*) was killed last year in this neighbourhood.

Our streams, which are small, and rise only at the end of the village, yield nothing but the bull's head or miller's thumb (*gobius fluviatilis capitatus*), the trout (*trutta fluviatilis*), the eel (*anguilla*), the lampern (*lampaetra parva et fluviatilis*), and the stickle-back (*pisciculus aculeatus*).[2]

We are twenty miles from the sea, and almost as many from a great river, and therefore see but little of sea birds. As to wild fowls, we have a few teams of ducks bred in the moors [marshes] where the snipes breed; and multitudes of widgeons and teals in hard weather frequent our lakes in the forest.

Having some acquaintance with a tame brown owl, I find that it casts up the fur of mice and the feathers of birds in pellets, after the manner of hawks; when full, like a dog, it hides what it cannot eat.

The young of the barn-owl are not easily raised, as they want a constant supply of fresh mice; whereas the young of the brown owl will eat indiscriminately all that is brought; snails, rats, kittens, puppies, magpies, and any kind of carrion or offal.

The house-martins have eggs still, and squab young. The last swift I observed was about the 21st of August: it was a straggler.

Red-starts, fly-catchers, white-throats, and *reguli non cristati*, still appear: but I have seen no blackcaps lately.

I forgot to mention that I once saw, in Christ Church College quadrangle in Oxford, on a very sunny warm morning, a house-martin flying about, and settling on the parapet, so late as the 20th of November.

At present I know only two species of bats, the common *vespertilio murinus* and the *vespertilio auribus*.[3]

I was much entertained last summer with a tame bat, which would take flies out of a person's hand. If you gave it anything to eat, it brought its wings round before the mouth, hovering and hiding its head in the manner of birds of prey when they feed. The adroitness

2 I do not attempt to identify the particular species here intended, in the absence of any sufficient description. ED.

3 It is not probable that White had seen the true *Vespertilio murinus*, which is a very rare bat; what he mistook for it must have been the Pipistrelle. His other species was doubtless the long-eared bat, *Plecotus auritus*. ED.

Long-eared bat

it showed in shearing off the wings of the flies, which were always rejected, was worthy of observation, and pleased me much. Insects seemed to be most acceptable, though it did not refuse raw flesh when offered; so that the notion, that bats go down chimneys and gnaw men's bacon, seems no improbable story. While I amused myself with this wonderful quadruped, I saw it several times confute the vulgar opinion, that bats when down upon a flat surface cannot get on the wing again, by rising with great ease from the floor. It ran, I observed, with more dispatch than I was aware of; but in a most ridiculous and grotesque manner.

Bats drink on the wing, like swallows, by sipping the surface, as they play over pools and streams. They love to frequent waters not only for the sake of drinking, but on account of insects, which are found over them in the greatest plenty. As I was going some years ago, pretty late, in a boat from Richmond to Sunbury, on a warm summer's evening, I think I saw myriads of bats between the two places; the air swarmed with them all along the Thames, so that hundreds were in sight at a time. I am, &c.

Letter 12 also to Thomas Pennant, Esq.

November 4th, 1767

Sir – It gave me no small satisfaction to hear that the *falco*[1] turned out an uncommon one. I must confess I should have been better pleased to have heard that I had sent you a bird that you had never seen before; but that, I find, would be a difficult task.

I have procured some of the mice[2] mentioned in my former letters, a young one and a female with young, both of which I have preserved in brandy. From the colour, shape, size and manner of nesting, I make no doubt but that the species is nondescript. They are much smaller, and more slender, than the *mus domesticus medius* of Ray; and have more of the squirrel or dormouse colour; their

1 This hawk proved to be the *falco peregrinus*; a variety.

2 Harvest-mice, which White was the first to discover and describe in England. ED.

belly is white, a straight line along their sides divides the shades of their back and belly. They never enter into houses; are carried into ricks and barns with the sheaves; abound in harvest; and build their nests amidst the straws of the corn above the ground, and sometimes in thistles. They breed as many as eight at a litter, in a little round nest composed of the blades of grass or wheat.

One of these nests I procured this autumn, most artificially platted, and composed of the blades of wheat, perfectly round, and about the size of a cricket-ball; with the aperture so ingeniously closed, that there was no discovering to what part it belonged. It was so compact and well filled, that it would roll across the table without being discomposed, though it contained eight little mice that were naked and blind. As this nest was perfectly full, how could the dam come at her litter respectively so as to administer a teat to each? Perhaps she opens different places for that purpose, adjusting them again when the business is over; but she could not possibly be contained herself in the ball with her young, which moreover would be daily increasing in bulk. This wonderful procreant cradle, an elegant instance of the efforts of instinct, was found in a wheat-field suspended in the head of a thistle.

A gentleman, curious in birds, wrote me word that his servant had shot one last January, in that severe weather, which he believed would puzzle me. I called to see it this summer, not knowing what to expect, but the moment I took it in hand, I pronounced it the male *garrulus bohemicus* or German silk-tail, from the five peculiar crimson tags or points which it carries at the ends of five of the short remiges. It cannot, I suppose, with any propriety, be called an English bird; and yet I see, by Ray's *Philosophical Letters*, that great flocks of them, feeding on haws, appeared in this kingdom in the winter of 1685.[3]

The mention of haws puts me in mind that there is a total failure of that wild fruit, so conducive to the support of many of the winged nation. For the same severe weather, late in the spring, which cut off all the produce of the more tender and curious trees, destroyed also that of the more hardy and common.

Some birds, haunting with the missel-thrushes, and feeding on the berries of the yew tree, which answered to the description of the *merula torquata*, or ring-ouzel, were lately seen in this neighbourhood.

3 The Bohemian wax-wing, an occasional visitor to England, appears at long intervals in considerable numbers. ED.

I employed some people to procure me a specimen, but without success. (See Letter 20.)

Query. – Might not canary birds be naturalised to this climate, provided their eggs were put, in the spring, into the nests of some of their congeners, as goldfinches, greenfinches, &c.? Before winter perhaps they might be hardened, and able to shift for themselves.

About ten years ago I used to spend some weeks yearly at Sunbury, which is one of those pleasant villages lying on the Thames, near Hampton Court. In the autumn, I could not help being much amused with those myriads of the swallow kind which assemble in those parts. But what struck me most was, that, from the time they began to congregate, forsaking the chimneys and houses, they roosted every night in the osier-beds of the aits [eyots] of that river. Now this resorting towards that element at that season of the year, seems to give some countenance to the northern opinion (strange as it is) of their retiring under water. A Swedish naturalist is so much persuaded of that fact, that he talks, in his calendar of Flora, as familiarly of the swallow's going under water in the beginning of September, as he would of his poultry going to roost a little before sunset.

Hawfinch

An observing gentleman in London writes me word that he saw an house-martin, on the twenty-third of last October, flying in and out of its nest in the Borough. And I myself, on the twenty-ninth of last October (as I was travelling through Oxford), saw four or five swallows hovering round and settling on the roof of the county hospital.

Now is it likely that these poor little birds (which perhaps had not been hatched but a few weeks) should, at that late season of the year, and from so midland a county, attempt a voyage to Goree or Senegal, almost as far as the equator?[4]

I acquiesce entirely in your opinion – that, though most of the swallow kind may migrate, yet that some do stay behind and hide with us during the winter.[5]

4 See Adanson's *Voyage to Senegal*.
5 This opinion is now known to be quite erroneous. ED.

As to the short-winged soft-billed birds, which come trooping in such numbers in the spring, I am at a loss even what to suspect about them. I watched them narrowly this year, and saw them abound till about Michaelmas, when they appeared no longer. Subsist they cannot openly among us, and yet elude the eyes of the inquisitive: and, as to their hiding, no man pretends to have found any of them in a torpid state in the winter. But with regard to their migration, what difficulties attend that supposition! that such feeble bad fliers (who the summer long never flit but from hedge to hedge) should be able to traverse vast seas and continents in order to enjoy milder seasons amidst the regions of Africa.

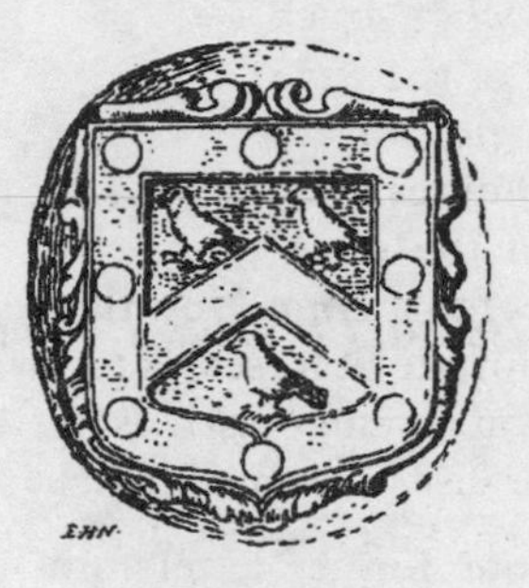

The White Arms

Hen Chaffinch

Letter 13 also to Thomas Pennant, Esq.

Selborne, January 22nd, 1768

Sir – As in one of your former letters you expressed the more satisfaction from my correspondence on account of my living in the most southerly county; so now I may return the compliment, and expect to have my curiosity gratified by your living much more to the north.

For many years past I have observed that towards Christmas vast flocks of chaffinches have appeared in the fields; many more, I used to think, than could be hatched in any one neighbourhood. But when I came to observe them more narrowly, I was amazed to find that they seemed to me to be almost all hens. I communicated my suspicions to some intelligent neighbours, who, after taking pains about the matter, declared that they also thought them all mostly females – at least fifty to one. This extraordinary occurrence brought to my mind the remark of Linnaeus; that 'before winter all their hen chaffinches migrate through Holland into Italy'. Now I want to know, from some curious person in the north, whether there are any large flocks of these finches with them in the winter, and of which sex they mostly consist? For, from such intelligence, one might be able to judge whether our female flocks migrate from the other end of the island, or whether they come over to us from the continent.

We have, in the winter, vast flocks of the common linnets: more, I think, than can be bred in any one district. These, I observe, when the spring advances, assemble on some tree in the sunshine, and join all in a gentle sort of chirping, as if they were about to break up their winter quarters and betake themselves to their proper summer homes. It is well known, at least, that the swallows and the fieldfares do congregate with a gentle twittering before they make their respective departure.

You may depend on it that the bunting, *Emberiza miliaria*, does not leave this county [Hampshire] in the winter. In January 1767, I saw several dozen of them, in the midst of a severe frost, among the bushes on the downs near Andover: in our woodland enclosed district it is a rare bird.

Wagtails, both white and yellow, are with us all the winter. Quails crowd to our southern coast, and are often killed in numbers by people that go on purpose.

Mr Stillingfleet, in his Tracts, says that 'if the wheatear (*œnanthe*) does not quit England, it certainly shifts places; for about harvest they are not to be found, where there was before great plenty of them.' This well accounts for the vast quantities that are caught about that time on the south downs near Lewes, where they are esteemed a delicacy. There have been shepherds, I have been credibly informed, that have made many pounds in a season by catching them in traps. And though such multitudes are taken, I never saw (and I am well acquainted with those parts) above two or three at a time, for they are never gregarious. They may perhaps migrate in general; and, for that purpose, draw towards the coast of Sussex in autumn: but that they do not all withdraw I am sure; because I see a few stragglers in many counties, at all times of the year, especially about warrens and stone quarries.

Wheatear

I have no acquaintance, at present, among the gentlemen of the navy; but have written to a friend, who was a sea-chaplain in the late war, desiring him to look into his minutes, with respect to birds that settled on their rigging during their voyage up or down the channel. What Hasselquist says on that subject is remarkable; there were little short-winged birds frequently coming on board his ship all the way from our channel quite up to the Levant, especially before squally weather.

What you suggest, with regard to Spain, is highly probable. The winters of Andalusia are so mild, that, in all likelihood, the soft-billed birds that leave us at that season may find insects sufficient to support them there.

Some young man, possessed of fortune, health, and leisure, should make an autumnal voyage into that kingdom; and should spend a year there, investigating the natural history of that vast country. Mr Willughby[1] passed through that kingdom on such an errand; but he seems to have skirted along in a superficial manner and an ill-humour, being much disgusted at the rude, dissolute manners of the people.

I have no friend left now at Sunbury to apply to about the swallows roosting on the aits of the Thames: nor can I hear any more about those birds which I suspected were*Merulae torquatae*.

As to the small mice [harvest-mice], I have farther to remark, that though they hang their nests for breeding up amidst the straws of the standing corn, above the ground; yet I find that, in the winter, they burrow deep in the earth, and make warm beds of grass: but their grand rendezvous seems to be in corn-ricks, into which they are carried at harvest. A neighbour housed an oat-rick lately, under the thatch of which were assembled near an hundred, most of which were taken, and some I saw. I measured them; and found that, from nose to tail, they were just two inches and a quarter, and their tails just two inches long. Two of them, in a scale, weighed down just one copper half-penny, which is about the third of an ounce avoirdupois: so that I suppose they are the smallest quadrupeds in this island. A full-grown *Mus medius domesticus* weighs, I find, one ounce lumping weight, which is more than six times as much as the mouse above; and measures from nose to rump four inches and a quarter, and the same in its tail.

We have had a very severe frost and deep snow this month. My thermometer was one day fourteen degrees and a half below the freezing-point, within doors. The tender evergreens were injured pretty much. It was very providential that the air was still, and the

1 See Ray's *Travels*, p. 466.

ground well covered with snow, else vegetation in general must have suffered prodigiously. There is reason to believe that some days were more severe than any since the year 1739-40.

I am, &c. &c.

Selborne & Nore Hill

Letter 14 also to Thomas Pennant, Esq.

Selborne, March 12th, 1768

Dear Sir – If some curious gentleman would procure the head of a fallow-deer, and have it dissected, he would find it furnished with two spiracula, or breathing-places, besides the nostrils; probably analogous to the *puncta lachrymalia* in the human head. When deer are thirsty, they plunge their noses, like some horses, very deep under water, while in the act of drinking, and continue them in that situation for a considerable time: but to obviate any inconveniency, they can open two vents, one at the inner corner of each eye, having a communication with the nose.[1] Here seems to be an extraordinary provision of nature worthy our attention; and which has not, that I

1 The glands to which White here alludes are not breathing-places, and have no connection with respiration. They are secretive organs, found also in all deer and antelopes, as well as in the common sheep: they exude an odorous body, which is probably important as adding to the attractiveness of its possessor, like musk and civet. It belongs to a group of special allurements the origin of which is fully worked up by Darwin in *The Descent of Man*. ED.

know of, been noticed by any naturalist. For it looks as if these creatures would not be suffocated, though both their mouths and nostrils were stopped. This curious formation of the head may be of singular service to beasts of chase, by affording them free respiration: and no doubt these additional nostrils are thrown open when they are hard run.[2] Mr Ray observed that at Malta the owners slit up the nostrils of such asses as were hard worked: for they, being naturally straight or small, did not admit air sufficient to serve them when they travelled, or laboured, in that hot climate. And we know that grooms, and gentlemen of the turf, think large nostrils necessary, and a perfection, in hunters and running horses.

Oppian, the Greek poet, by the following line, seems to have had some notion that stags have four spiracula:

Τετραδύμοι ῥῖνες, πίρυρες πνοιῇσι δίαυλοι.

Quadrifidae nares, quadruplices ad respirationem canales.

OPPIAN, *Cynegetica*, Book ii, line 181

Writers, copying from one another, make Aristotle say that goats breathe at their ears: whereas he asserts just the contrary:

Αλκμαίων γὰρ ὄυκ ἀληυὴ λέγει, φάμενος ἀνάπνειν τάς αἰγὰς κατὰ τὰ ὦτα.

Alcmaeon does not advance what is true, when he avers that goats breathe through their ears.

History of Animals, Book I, chapter xi

2 In answer to this account, Mr Pennant sent me the following curious and pertinent reply: 'I was much surprised to find in the antelope something analogous to what you mention as so remarkable in deer. This animal also has a long slit beneath each eye, which can be opened and shut at pleasure. On holding an orange to one, the creature made as much use of those orifices as of his nostrils, applying them to the fruit, and seeming to smell it through them.'

The church from the N·W·

Bullfinch

Letter 15 also to Thomas Pennant, Esq.

Selborne, March 30th, 1768

Dear Sir – Some intelligent country people have a notion that we have, in these parts, a species of the *genus mustelinum*, besides the weasel, stoat, ferret, and polecat; a little reddish beast, not much bigger than a field-mouse, but much longer, which they call a*cane*. This piece of intelligence can be little depended on; but farther inquiry may be made.[1]

A gentleman in this neighbourhood had two milk-white rooks in one nest. A booby of a carter, finding them before they were able to fly, threw them down and destroyed them, to the regret of the owner, who would have been glad to have preserved such a curiosity in his rookery. I saw the birds myself nailed against the end of a barn, and was surprised to find that their bills, legs, feet, and claws were milk-white.

A shepherd saw, as he thought, some white larks on a down above my house this winter: were not these the *Emberiza nivalis*, the

1 There is no such animal known to science in Britain, though gamekeepers and others still stoutly assert its existence in many places. Female weasels and the young, when attempting to escape, have a habit of shrinking into themselves, so as to look very small – a peculiarity which doubtless has given rise to the persistent delusion. ED.

snowflake of the *British Zoology*? No doubt they were.

A few years ago I saw a cock bullfinch in a cage which had been caught in the fields after it was come to its full colours. In about a year it began to look dingy; and, blackening every succeeding year, it became coal black at the end of four. Its chief food was hempseed. Such influence has food on the colour of animals! The pied and mottled colours of domesticated animals are supposed to be owing to high, various, and unusual food.

I had remarked, for years, that the root of the cuckoo-pint (*arum*) was frequently scratched out of the dry banks of hedges, and eaten in severe snowy weather. After observing, with some exactness, myself, and getting others to do the same, we found it was the thrush kind that searched it out. The root of the *arum is* remarkably warm and pungent.

Our flocks of female chaffinches have not yet forsaken us. The blackbirds and thrushes are very much thinned down by that fierce weather in January.

In the middle of February I discovered, in my tall hedges, a little bird that raised my curiosity: it was of that yellow-green colour that belongs to the *salicaria* kind, and, I think, was soft-billed. It was no *parus*; and was too long and too big for the golden-crowned wren, appearing most like the largest willow-wren. It hung sometimes with its back downwards, but never continuing one moment in the same place. I shot at it, but it was so desultory that I missed my aim.

I wonder that the stone-curlew, *Charadrius oedicnemus*, should be mentioned by the writers as a rare bird: it abounds in all the champaign parts of Hampshire and Sussex, and breeds, I think, all the summer, having young ones, I know, very late in the autumn. Already they begin clamouring in the evening. They cannot, I think, with any propriety, be called, as they are by Mr Ray, '*circa aquas versantes*'; for with us, by day at least, they haunt only the most dry, open, upland fields and sheep-walks, far removed from water: what they may do in the night I cannot say. Worms are their usual food, but they also eat toads and frogs.[2]

I can show you some good specimens of my new mice. Linnaeus perhaps would call the species *Mus minimus*.

2 White was mistaken in supposing that the stone-curlew does not frequent water. He knew it only in its summer breeding-time. In winter it habitually haunts wet and marshy places. ED.

Stone-curlew

Letter 16 also to Thomas Pennant, Esq.

Selborne, April 18th, 1768

Dear Sir – The history of the stone-curlew, *Charadrius oedicnemus*, is as follows.[1] It lays its eggs, usually two, never more than three, on the bare ground, without any nest, in the field; so that the countryman, in stirring his fallows, often destroys them. The young run immediately from the egg, like partridges, &c., and are withdrawn to some flinty field by the dam, where they sculk among the stones, which are their best security; for their feathers are so exactly of the colour of our grey spotted flints, that the most exact observer, unless he catches the eye of the young bird, may be eluded. The eggs are short and round; of a dirty white, spotted with dark bloody

1 These remarks are obviously called out by a question from Pennant respecting the stone-curlew mentioned in the last letter. ED.

Chiffchaff

blotches. Though I might not be able, just when I pleased, to procure you a bird, yet I could show you them almost any day; and any evening you may hear them round the village, for they make a clamour which may be heard a mile. *Oedicnemus* is a most apt and expressive name for them, since their legs seem swollen like those of a gouty man. After harvest I have shot them before the pointers in turnip fields.

I make no doubt but there are three species of the willow-wrens;[2] two I know perfectly, but have not been able yet to procure the third. No two birds can differ more in their notes, and that constantly, than those two that I am acquainted with; for the one has a joyous, easy, laughing note, the other a harsh, loud chirp. The former is every way larger, and three-quarters of an inch longer, and weighs two drams and a half, while the latter weighs but two; so the songster is one-fifth heavier than the chirper. The chirper (being the first summer-bird of passage that is heard, the wryneck sometimes excepted) begins his two notes in the middle of March, and continues them through the spring and summer till the end of August, as appears by my journals. The legs of the larger of these two are flesh-coloured; of the less black.

The grasshopper-lark[3] began his sibilous note in my fields last Saturday. Nothing can be more amusing than the whisper of this little bird, which seems to be close by though at a hundred yards distance; and, when close at your ear, is scarce any louder than when a great way off. Had I not been a little acquainted with insects, and known that the grasshopper kind is not yet hatched, I should have

Willowwren

2 These are doubtless the wood-wren, *Sylvia* (*Phylloscopus*) *sibilatrix*, called by White the songster; the willow-wren, *Sylvia* (*Phylloscopus*) *trochilus*; and the chiffchaff, *Sylvia hippolais* (or *Phylloscopus rufus*), called by White the chirper. ED.

3 Now called the grasshopper-warbler, *Salicaria locustella* (*Locustella naevia*). ED.

hardly believed but that it had been a *locusta* whispering in the bushes. The country people laugh when you tell them that it is the note of a bird. It is a most artful creature, skulking in the thickest part of a bush; and will sing at a yard distance, provided it be concealed. I was obliged to get a person to go on the other side of the hedge where it haunted, and then it would run, creeping like a mouse, before us for a hundred yards together, through the bottom of the thorns; yet it would not come into fair sight; but in a morning, early, and when undisturbed, it sings on the top of a twig, gaping and shivering with its wings. Mr Ray himself had no knowledge of this bird, but received his account from Mr Johnson, who apparently confounds it with the *reguli non cristati*, from which it is very distinct. See Ray's *Philosophical Letters*, page 108.

A LIST OF THE SUMMER BIRDS OF PASSAGE DISCOVERED IN THIS NEIGHBOURHOOD, RANGED SOMEWHAT IN THE ORDER IN WHICH THEY APPEAR

	LINNAEI NOMINA
Smallest willow-wren	*Motacilla trochilus*
Wryneck	*Jynx torquilla*
House-swallow	*Hirundo rustica*
Martin	*Hirundo urbica*
Sand-martin	*Hirundo riparia*
Cuckoo	*Cuculus canorus*
Nightingale	*Motacilla luscinia*
Blackcap	*Motacilla atricapilla*
Whitethroat	*Motacilla sylvia*
Middle willow-wren	*Motacilla trochilus*
Swift	*Hirundo apus*
Stone-curlew (?)	*Charadrius oedicnemus* (?)
Turtle-dove (?)	*Turtur aldrovandi* (?)
Grasshopper-lark	*Alauda trivialis*
Landrail	*Rallus crex*
Largest willow-wren	*Motacilla trochilus*
Redstart	*Motacilla phaenicurus*
Goat-sucker, or fern-owl	*Caprimulgus europaeus*
Fly-catcher	*Muscicapa grisola*

The fly-catcher (*stoparola*) has not yet appeared; it usually breeds in my vine. The redstart begins to sing; its note is short and imperfect, but is continued till about the middle of June. The willow wrens (the smaller sort) are horrid pests in a garden,

Grasshopper-warbler

destroying the peas, cherries, currants, &c.; and are so tame that a gun will not scare them.

My countrymen talk much of a bird that makes a clatter with its bill against a dead bough, or some old pales, calling it a jarbird. I procured one to be shot in the very fact; it proved to be the *Sitta europaea* (the nuthatch). Mr Ray says that the less spotted woodpecker does the same. This noise may be heard a furlong or more.

Now is the only time to ascertain the short-winged summer birds; for, when the leaf is out, there is no making any remarks on such a restless tribe; and, when once the young begin to appear, it is all confusion: there is no distinction of genus, species, or sex.

In breeding-time snipes play over the moors, piping and humming: they always hum as they are descending. Is not their hum ventriloquous like that of the turkey? Some suspect it is made by their wings.

This morning I saw the golden-crowned wren, whose crown glitters like burnished gold. It often hangs like a titmouse, with its back downwards.

Yours, &c. &c.

Golden-crested wren

The Wakes

Letter 17 also to Thomas Pennant, Esq.

Selborne, June 18th, 1768

Dear Sir – On Wednesday last arrived your agreeable letter of June the 10th. It gives me great satisfaction to find that you pursue these studies still with such vigour, and are in such forwardness with regard to reptiles and fishes.

The reptiles, few as they are, I am not acquainted with, so well as I could wish, with regard to their natural history. There is a degree of dubiousness and obscurity attending the propagation of this class of animals, something analogous to that of the *cryptogamia* in the sexual system of plants: and the case is the same with regard to some of the fishes; as the eel, &c.

The method in which toads procreate and bring forth seems to be very much in the dark. Some authors say that they are viviparous:

and yet Ray classes them among his oviparous animals; and is silent with regard to the manner of their bringing forth. Perhaps they may be ἔσω μὲν ᾠοτόκοι, ἔξω δε ζωοτόκοι, as is known to be the case with the viper.

The copulation of frogs (or at least the appearance of it; for Swammerdam proves that the male has no *penis intrans*) is notorious to everybody: because we see them sticking upon each other's backs for a month together in the spring: and yet I never saw or read of toads being observed in the same situation. It is strange that the matter with regard to the venom of toads has not been yet settled. That they are not noxious to some animals is plain: for ducks, buzzards, owls, stone-curlews, and snakes, eat them, to my knowledge, with impunity. And I well remember the time, but was not eye-witness to the fact (though numbers of persons were), when a quack, at this village, ate a toad to make the country-people stare; afterwards he drank oil.[1]

I have been informed also, from undoubted authority, that some ladies (ladies you will say of peculiar taste) took a fancy to a toad, which they nourished summer after summer, for many years, till he grew to a monstrous size, with the maggots which turn to flesh-flies. The reptile used to come forth every evening from a hole under the garden-steps; and was taken up, after supper, on the table to be fed. But at last a tame raven, kenning him as he put forth his head, gave him such a severe stroke with his horny beak as put out one eye. After this accident the creature languished for some time and died.

I need not remind a gentleman of your extensive reading of the excellent account there is from Mr Derham, in Ray's *Wisdom of God in the Creation* (p. 365), concerning the migration of frogs from their breeding ponds. In this account he at once subverts that foolish opinion of their dropping from the clouds in rain; showing that it is from the grateful coolness and moisture of those showers that they are tempted to set out on their travels, which they defer till those fall. Frogs are as yet in their tadpole state; but, in a few weeks, our lanes, paths, fields, will swarm for a few days with myriads of those emigrants, no larger than my little finger nail.

1 There is no venom in toads, though they have a nasty taste which makes dogs and fish reject them. The almost universal notion that they are poisonous or dangerous is a pure superstition, which originated, no doubt, in their repulsive appearance. ED.

Dorton Cottage

Swammerdam gives a most accurate account of the method and situation in which the male impregnates the spawn of the female. How wonderful is the economy of Providence with regard to the limbs of so vile a reptile! While it is an aquatic it has a fishlike tail, and no legs: so soon as the legs sprout, the tail drops of[2] as useless, and the animal betakes itself to the land!

Merret, I trust, is widely mistaken when he advances that the *Rana arborea* is an English reptile; it abounds in Germany and Switzerland.

It is to be remembered that the *Salamandra aquatica* of Ray (the water-newt or eft) will frequently bite at the angler's bait, and is often caught on his hook. I used to take it for granted that the *Salamandra aquatica* was hatched, lived, and died, in the water. But John Ellis, Esq., F.R.S. (the coralline Ellis), asserts, in a letter to the Royal Society, dated June the 5th, 1766, in his account of the *mud inguana*, an amphibious bipes from South Carolina, that the water-eft, or newt, is only the larva of the land-eft, as tadpoles are of frogs. Lest I should be suspected to misunderstand his meaning, I shall give it in his own words. Speaking of the *opercula* or coverings to the gills of the *mud inguana*, he proceeds to say that, 'The form of these pennated coverings approaches very near to what I have some time ago observed in the larva or aquatic state of our English *lacerta*, known by the name of eft, or newt; which serve them for coverings to their gills, and for fins to swim with while in this state; and which they lose, as well as the fins of their tails, when they change their state and become land animals, as I have observed, by keeping them alive for some time myself.'[3]

Linnaeus, in his *Systema Naturae*, hints at what Mr Ellis advances more than once.

Providence has been so indulgent to us as to allow of but one venomous reptile of the serpent kind in these kingdoms, and that is the viper. As you propose the good of mankind to be an object of your publications, you will not omit to mention common salad oil as a sovereign remedy against the bite of the viper. As to the blind worm (*Anguis fragilis*, so called because it snaps in sunder with a

2 Or, rather, is absorbed. ED.

3 The newt is not the larva of the lizard, but is an amphibian belonging to a distinct order. It passes, however, through a tadpole state, during which it is provided with fish-like gills. In White's time, the relations of the two were ill understood. For particulars, see any good modern work on natural history. ED.

Common snake

small blow), I have found, on examination, that it is perfectly innocuous. A neighbouring yeoman (to whom I am indebted for some good hints) killed and opened a female viper about the 27th of May: he found her filled with a chain of eleven eggs, about the size of those of a blackbird; but none of them were advanced so far towards a state of maturity as to contain any rudiments of young. Though they are oviparous, yet they are viviparous also, hatching their young within their bellies, and then bringing them forth. Whereas snakes[4] lay chains of eggs every summer in my melon beds, in spite of all that my people can do to prevent them; which eggs do not hatch till the spring following, as I have often experienced. Several intelligent folks assure me that they have seen the viper open her mouth, and admit her helpless young down her throat on sudden surprises, just as the female opossum does her brood into the pouch under her belly, upon the like emergencies; and yet the London viper-catchers insist on it, to Mr Barrington, that no such thing ever happens.[5] The serpent kind eat, I believe, but once in a year; or rather, but only just at one season of the year. Country people talk much of a water-snake, but, I am pretty sure, without any reason; for the common snake (*Coluber natrix*) delights much to sport in the water, perhaps with a view to procure frogs and other food.

I cannot well guess how you are to make out your twelve species of reptiles, unless it be by the various species, or rather varieties, of our *lacerti*, of which Ray enumerates five. I have not had opportunity of ascertaining these; but remember well to have seen, formerly, several beautiful green *lacerti* on the sunny sandbanks near Farnham, in Surrey; and Ray admits there are such in Ireland.

4 That is to say, ring-snakes, as opposed to vipers. ED.

5 This question is not even now absolutely settled, though I do not doubt the story is a pure medieval superstition. ED.

The Hanger from Dorton

Letter 18 also to Thomas Pennant, Esq.

Selborne, July 27th, 1768

Dear Sir – I received your obliging and communicative letter of June the 28th, while I was on a visit at a gentleman's house, where I had neither books to turn to, nor leisure to sit down, to return you an answer to many queries, which I wanted to resolve in the best manner that I am able.

A person, by my order, has searched our brooks, but could find no such fish as the *Gasterosteus pungitius*: he found the *Gasterosteus aculeatus* in plenty. This morning, in a basket, I packed a little earthen pot full of wet moss, and in it some sticklebacks, male and female; the females big with spawn: some lamperns; some bull's heads; but I could procure no minnows. This basket will be in Fleet Street by eight this evening; so I hope Mazel[1] will have them fresh and fair tomorrow morning. I gave some directions, in a letter, to what particulars the engraver should be attentive.

1 Pennant's engraver. ED.

Finding, while I was on a visit, that I was within a reasonable distance of Ambresbury [Amesbury], I sent a servant over to that town, and procured several living specimens of loaches, which he brought, safe and brisk, in a glass decanter. They were taken in the gullies that were cut for watering the meadows. From these fishes (which measured from two to four inches in length) I took the following description: 'The loach, in its general aspect, has a pellucid appearance; its back is mottled with irregular collections of small black dots, not reaching much below the *linea lateralis*, as are the back and tail fins; a black line runs from each eye down to the nose; its belly is of a silvery white; the upper jaw projects beyond the lower, and is surrounded with six feelers, three on each side; its pectoral fins are large, its ventral much smaller; the fin behind its anus small; its dorsal fin large, containing eight spines; its tail where it joins to the tail-fin, remarkably broad, without any taperness, so as to be characteristic of this genus; the tail-fin is broad, and square at the end. From the breadth and muscular strength of the tail it appears to be an active, nimble fish.'

In my visit I was not very far from Hungerford, and did not forget to make some inquiries concerning the wonderful method of curing cancers by means of toads. Several intelligent persons, both gentry and clergy, do, I find, give a great deal of credit to what is asserted in the papers, and I myself dined with a clergyman who seemed to be persuaded that what is related is matter of fact; but, when I came to attend to his account, I thought I discerned circumstances which did not a little invalidate the woman's story of the manner in which she came by her skill. She says of herself 'that, labouring under a virulent cancer, she went to some church where there was a vast crowd; on going into a pew, she was accosted by a strange clergyman, who, after expressing compassion for her situation, told her that if she would make such an application of living toads as is mentioned she would be well.' Now is it likely that this unknown gentleman should express so much tenderness for this single sufferer, and not feel any for the many thousands that daily languish under this terrible disorder? Would he not have made use of this invaluable nostrum for his own emolument; or at least, by some means of publication or other, have found a method of making it public for the good of mankind? In short, this woman (as it appears to me), having set up for a cancer-doctress, finds it expedient to amuse the country with this dark and mysterious relation.

The water-eft has not, that I can discern, the least appearance of any gills; for want of which it is continually rising to the surface of

The church from the S.E.

the water to take in fresh air.[2] opened a big-bellied one indeed, and found it full of spawn. Not that this circumstance at all invalidates the assertion that they are *larvae*; for the *larvae* of insects are full of eggs, which they exclude the instant they enter their last state. The water-eft is continually climbing over the brims of the vessel, within which we keep it in water, and wandering away; and people every summer see numbers crawling out of the pools where they are hatched up the dry banks. There are varieties of them differing in colour; and some have fins up their tail and back, and some have not.[3]

2 White is quite right as to the newt in its developed adult state: but in its larval form, Ellis was correct in saying that it possesses gills. ED.

3 The male newts develop an ornamental jagged crest or membrane up the tail and back in the breeding season only, doubtless as an attraction to add to their beauty. ED

In Selborne Street

Letter 4 also to Thomas Pennant, Esq.

Selborne, August 17th, 1768

Dear Sir – I have now, past dispute, made out three distinct species of the willow-wrens (*motacillae trochili*) which constantly and invariably use distinct notes. But at the same time I am obliged to confess that I know nothing of your willow-lark.[1] In my letter of April the 18th, I had told you peremptorily that I knew your willow-lark, but had not seen it then; but when I came to procure it, it proved in all respects a very *motacilla trochilus*, only that it is a size larger than the two other, and the yellow-green of the whole upper part of the body is more vivid, and the belly of a clearer white. I have specimens of the three sorts now lying before me, and can discern that there are three

Willow-wren

1 *British Zoology*, 1776, p. 381.

gradations of sizes, and that the least has black legs, and the other two flesh-coloured ones. The yellowest bird is considerably the largest, and has its quill-feathers and secondary feathers tipped with white, which the others have not. This last haunts only the tops of trees in high beechen woods, and makes a sibilous, grasshopper-like noise, now and then, at short intervals, shivering a little with its wings when it sings; and is, I make no doubt now, the *regulus non cristatus* of Ray, which he says, '*cantat voce stridula locustae*'. Yet this great ornithologist never suspected that there were three species.

Wood-wren

Ring-ousel

Letter 20 also to Thomas Pennant, Esq.

Selborne, October 8th, 1768

It is I find in zoology as it is in botany; all nature is so full that that district produces the greatest variety which is the most examined. Several birds, which are said to belong to the north only, are it seems often in the south. I have discovered this summer three species of birds with us, which writers mention as only to be seen in the northern counties. The first that was brought me (on the 14th of May) was the sandpiper, *tringa hypoleucus*: it was a cock bird, and haunted the banks of some ponds near the village; and, as it had a companion, doubtless intended to have bred near that water.

Besides, the owner has told me since, that on recollection, he has seen some of the same birds round his ponds in former summers![1]

The next bird that I procured (on the 21st of May) was a male red-backed butcher-bird, *lanius collurio*. My neighbour, who shot it, says that it might easily have escaped his notice, had not the outcries

1 The sandpiper is not, as White thought, a specially northern bird. It occurs in most parts of our southern counties. ED.

and chattering of the whitethroats and other small birds drawn his attention to the bush where it was; its craw was filled with the legs and wings of beetles. The next rare birds (which were procured for me last week) were some ring-ousels, *turdi torquati*.

This week twelve months a gentleman from London, being with us, was amusing himself with a gun, and found, he told us, on an old yew hedge where there were berries some birds like blackbirds, with rings of white round their necks: a neighbouring farmer also at the same time observed the same; but, as no specimens were procured, little notice was taken. I mentioned this circumstance to you in my letter of November the 4th, 1767 [Letter 12] (you, however, paid but small regard to what I said, as I had not seen these birds myself); but last week the aforesaid farmer, seeing a large flock, twenty or thirty of these birds, shot two cocks and two hens, and says, on recollection, that he remembers to have observed these birds again last spring, about Lady-day, as it were on their return to the north. Now perhaps these ousels are not the ousels of the north of England, but belong to the more northern parts of Europe; and may retire before the excessive rigour of the frosts in those parts, and return to breed in the spring, when the cold abates. If this be the case, here is discovered a new bird of winter passage, concerning whose migrations the writers are silent; but if these birds should prove the ousels of the north of England, then here is a migration disclosed within our own kingdom never before remarked. It does not yet appear whether they retire beyond the bounds of our island to the south; but it is most probable that they usually do, or else one cannot suppose that they would have continued so long unnoticed in the southern countries. The ousel is larger than a blackbird, and feeds on haws; but last autumn (when there were no haws) it fed on yew-berries: in the spring it feeds on ivy-berries, which ripen only at that season, in March and April.

I must not omit to tell you (as you have been so lately on the study of reptiles) that my people, every now and then of late, draw up with a bucket of water from my well, which is sixty-three feet deep, a large black warty lizard[2] with a fin-tail and yellow belly. How they first came down at that depth, and how they were ever to have got out thence without help, is more than I am able to say.

My thanks are due to you for your trouble and care in the examination of a buck's head. As far as your discoveries reach at

2 This was, no doubt, a Great Crested Newt. ED.

present, they seem much to corroborate my suspicions; and I hope Mr — may find reason to give his decision in my favour; and then, I think, we may advance this extraordinary provision of nature as a new instance of the wisdom of God in the creation.

As yet I have not quite done with my history of the *oedicnemus*, or stone-curlew; for I shall desire a gentleman in Sussex (near whose house these birds congregate in vast flocks in the autumn) to observe nicely when they leave him (if they do leave him), and when they return again in the spring: I was with this gentleman lately, and saw several single birds.

White's stool

Jackdaw

Letter 21 also to Thomas Pennant, Esq.

Selborne, November 28th, 1768

Dear Sir – With regard to the *oedicnemus*, or stone-curlew, I intend to write very soon to my friend near Chichester, in whose neighbourhood these birds seem most to abound; and shall urge him to take particular notice when they begin to congregate, and afterwards to watch them most narrowly whether they do not withdraw themselves during the dead of the winter. When I have obtained information with respect to this circumstance, I shall have finished my history of the stone-curlew; which I hope will prove to your satisfaction, as it will be, I trust, very near the truth. This gentleman, as he occupies a large farm of his own, and is abroad early and late, will be a very proper spy upon the motions of these birds; and besides, as I have prevailed on him to buy the *Naturalist's Journal* (with which he is much delighted), I shall expect that he will be very exact in his dates. It is very extraordinary, as you observe, that a bird so common with us should never straggle to you.

And here will be the properest place to mention, while I think of it, an anecdote which the above-mentioned gentleman told me when I was last at his house; which was that, in a warren joining to his outlet, many daws (*corvi monedulae*) build every year in the rabbit burrows under-ground. The way he and his brothers used to take their nests, while they were boys, was by listening at the mouths of the holes; and, if they heard the young ones cry, they twisted the nest out with a forked stick. Some water-fowls (viz., the puffins) breed, I know, in that manner; but I should never have suspected the daws of building in holes on the flat ground.

Another very unlikely spot is made use of by daws as a place to breed in, and that is Stonehenge. These birds deposit their nests in the interstices between the upright and the impost stones of that amazing work of antiquity: which circumstance alone speaks the prodigious height of the upright stones, that they should be tall enough to secure those nests from the annoyance of shepherd-boys, who are always idling round that place.

One of my neighbours last Saturday, November the 26th, saw a martin in a sheltered bottom: the sun shone warm, and the bird was hawking briskly after flies. I am now perfectly satisfied that they do not all leave this island in the winter.[1]

You judge very right, I think, in speaking with reserve and caution concerning the cures done by toads: for, let people advance what they will on such subjects, yet there is such a propensity in mankind towards deceiving and being deceived, that one cannot safely relate anything from common report, especially in print, without expressing some degree of doubt and suspicion.

Your approbation, with regard to my new discovery of the migration of the ring-ousel, gives me satisfaction; and I find you concur with me in suspecting that they are foreign birds which visit us. You will be sure, I hope, not to omit to make inquiry whether your ring-ousels leave your rocks in the autumn. What puzzles me most, is the very short stay they make with us; for in about three weeks they are all gone. I shall be very curious to remark whether they will call on us at their return in the spring, as they did last year.

I want to be better informed with regard to ichthyology. If fortune had settled me near the sea-side, or near some great river, my natural propensity would soon have urged me to have made myself acquainted with their productions: but as I have lived mostly in inland parts, and in an upland district, my knowledge of fishes extends little farther than to those common sorts which our brooks and lakes produce.

I am, &c.

1 In this our author was, of course, mistaken. ED.

Letter 22 also to Thomas Pennant, Esq.

Selborne, January 2nd, 1769

Dear Sir – As to the peculiarity of jackdaws building with us under the ground in rabbit burrows, you have, in part, hit upon the reason; for, in reality, there are hardly any towers or steeples in all this country. And perhaps, Norfolk excepted, Hampshire and Sussex are as meanly furnished with churches as almost any counties in the kingdom. We have many livings of two or three hundred pounds a year, whose houses of worship make little better appearance than dovecots. When I first saw Northamptonshire, Cambridgeshire, and Huntingdonshire, and the fens of Lincolnshire, I was amazed at the number of spires which presented themselves in every point of view. As an admirer of prospects, I have reason to lament this want in my own country; for such objects are very necessary ingredients in an elegant landscape.

What you mention with respect to reclaimed toads raises my curiosity. An ancient author, though no naturalist, has well remarked that 'every kind of beasts, and of birds, and of serpents, and things in the sea, is tamed, and hath been tamed, of mankind'.[1]

1 James, chap. iii. 7.

It is a satisfaction to me to find that a green lizard has actually been procured for you in Devonshire; because it corroborates my discovery, which I made many years ago, of the same sort, on a sunny sandbank near Farnham, in Surrey. I am well acquainted with the South Hams of Devonshire; and can suppose that district, from its southerly situation, to be a proper habitation for such animals in their best colours.

Since the ring-ousels of your vast mountains do certainly not forsake them against winter, our suspicions that those which visit this neighbourhood about Michaelmas are not English birds, but driven from the more northern parts of Europe by the frosts, are still more reasonable; and it will be worth your pains to endeavour to trace from whence they come, and to inquire why they make so very short a stay.

Heron

In your account of your error with regard to the two species of herons, you incidentally gave me great entertainment in your description of the heronry at Cressi Hall; which is a curiosity I never could manage to see. Fourscore nests of such a bird on one tree is a rarity which I would ride half as many miles to have a sight of. Pray be sure to tell me in your next whose seat Cressi Hall is, and near what town it lies.[2] I have often thought that those vast extents of fens have never been sufficiently explored. If half a dozen gentlemen, furnished with a good strength of water-spaniels, were to beat them over a week, they would certainly find more species.

There is no bird, I believe, whose manners I have studied more than that of the *caprimulgus* (the goat-suckers),[3] as it is a wonderful and curious creature; but I have always found that though sometimes it may chatter as it flies, as I know it does, yet in general it utters its jarring note sitting on a bough; and I have for many a half-hour watched it as it sat with its under mandible quivering, and particularly this summer. It perches usually on a bare twig, with its

2 Cressi Hall is near Spalding, in Lincolnshire.

3 More commonly known nowadays as the nightjar. ED.

head lower than its tail, in an attitude well expressed by your draughtsman in the folio *British Zoology*. This bird is most punctual in beginning its song exactly at the close of day; so exactly that I have known it strike up more than once or twice just at the report of the Portsmouth evening gun, which we can hear when the weather is still. It appears to me past all doubt that its notes are formed by organic impulse, by the powers of the parts of its windpipe, formed for sound, just as cats purr. You will credit me, I hope, when I assure you that, as my neighbours were assembled in an hermitage on the side of a steep hill where we drink tea, one of these churn-owls came and settled on the cross of that little straw edifice and began to chatter, and continued his note for many minutes; and we were all struck with wonder to find that the organs of that little animal, when put in motion, gave a sensible vibration to the whole building! This bird also sometimes makes a small squeak, repeated four or five times; and I have observed that to happen when the cock has been pursuing the hen in a toying way through the boughs of a tree.

It would not be at all strange if your bat, which you have procured, should prove a new one, since five species have been found in a neighbouring kingdom. The great sort that I mentioned is certainly a nondescript; I saw but one this summer, and that I had no opportunity of taking.

Your account of the Indian grass was entertaining. I am no angler myself; but inquiring of those that are, what they supposed that part of their tackle to be made of? – they replied, 'Of the intestines of a silkworm.'

Though I must not pretend to great skill in entomology, yet I cannot say that I am ignorant of that kind of knowledge; I may now and then perhaps be able to furnish you with a little information.

The vast rains ceased with us much about the same time as with you, and since we have had delicate weather. Mr Barker, who has measured the rain for more than thirty years, says, in a late letter, that more has fallen this year than in any he ever attended to; though from July 1763 to January 1764 more fell than in any seven months of this year.

Letter 23 also to Thomas Pennant, Esq.

Selborne, February 28th, 1769

Dear Sir – It is not improbable that the Guernsey lizard and our green lizards may be specifically the same; all that I know is, that, when some years ago many Guernsey lizards were turned loose in Pembroke College garden, in the University of Oxford, they lived a great while, and seemed to enjoy themselves very well, but never bred. Whether this circumstance will prove anything either way I shall not pretend to say.

I return you thanks for your account of Cressi Hall; but recollect, not without regret, that in June 1746 I was visiting for a week together at Spalding, without ever being told that such a curiosity was just at hand. Pray send me word in your next what sort of tree it is that contains such a quantity of herons' nests; and whether the heronry consists of a whole grove of wood, or only of a few trees.

It gave me satisfaction to find we accorded so well about the *caprimulgus*; all I contended for was to prove that it often chatters sitting as well as flying; and therefore the noise was voluntary, and from organic impulse, and not from the resistance of the air against the hollow of its mouth and throat.[1]

If ever I saw anything like actual migration, it was last Michaelmas Day. I was travelling, and out early in the morning; at first there was a vast fog; but, by the time that I was got seven or eight miles from home towards the coast, the sun broke out into a delicate warm day. We were then on a large heath or common, and I could discern, as the mist began to break away, great numbers of swallows (*hirundines rusticae*) clustering on the stunted shrubs and bushes, as if they had roosted there all night. As soon as the air became clear and pleasant they all were on the wing at once; and, by a placid and easy flight, proceeded on southward towards the sea; after this I did not see any more flocks, only now and then a straggler.

I cannot agree with those persons that assert that the swallow kind disappear some and some, gradually, as they come, for the bulk of them seem to withdraw at once; only some stragglers stay behind a long while, and do never, there is the greatest reason to believe, leave this island.[2] Swallows seem to lay themselves up, and to come forth in a warm day, as bats do continually of a warm evening, after they have disappeared for weeks. For a very respectable gentleman assured me that, as he was walking with some friends under Merton Wall[3] on a remarkably hot noon, either in the last week in December or the first week in January, he espied three or four swallows huddled together on the moulding of one of the windows of that college. I have frequently remarked that swallows are seen later at Oxford than elsewhere; is it owing to the vast massy buildings of that place, to the many waters round it, or to what else?

When I used to rise in a morning last autumn, and see the swallows and martins clustering on the chimneys and thatch of the neighbouring cottages, I could not help being touched with a secret delight, mixed with some degree of mortification; with delight, to observe with how much ardour and punctuality those poor little birds obeyed the strong impulse towards migration, or hiding,[4]

1 The night-jar usually churrs when seated (lengthwise) on a bough; the trilled noise is undoubtedly voluntary, and is a love-call to its mate. ED.

2 No swallows winter in England, though a few stragglers may be seen on warm days in late autumn or early spring. ED.

3 The wall between Merton College, Oxford, and Christ Church Meadows, familiar both to White and his correspondent. ED.

4 The idea of hiding is, of course, erroneous. ED.

The Lythe

imprinted on their minds by their great Creator; and with some degree of mortification, when I reflected that, after all our pains and inquiries, we are yet not quite certain to what regions they do migrate; and are still farther embarrassed to find that some do not actually migrate at all.

These reflections made so strong an impression on my imagination, that they became productive of a composition that may perhaps amuse you for a quarter of an hour when next I have the honour of writing to you.

Sedge-warbler

Letter 24 also to Thomas Pennant, Esq.

Selborne, May 29th, 1769

Dear Sir – The *scarabaeus fullo* I know very well, having seen it in collections; but have never been able to discover one wild in its natural state. Mr Banks told me he thought it might be found on the sea-coast.

On the thirteenth of April I went to the sheep-down, where the ring-ousels have been observed to make their appearance at spring and fall, in their way perhaps to the north or south; and was much pleased to see these birds about the usual spot. We shot a cock and a hen; they were plump and in high condition. The hen had but very small rudiments of eggs within her, which proves they are late breeders; whereas those species of the thrush kind that remain with us the whole year have fledged young before that time. In their crops was nothing very distinguishable, but somewhat that seemed like blades of vegetables nearly digested. In autumn they feed on haws and yew-berries, and in the spring on ivy-berries. I dressed one of these birds, and found it juicy and well flavoured. It is remarkable that they make but a few days' stay in their spring visit, but rest near a fortnight at Michaelmas. These birds, from the observations of three springs and two autumns, are most punctual in their return; and exhibit a new migration unnoticed by the writers, who supposed they never were to be seen in any southern countries.

One of my neighbours lately brought me a new*salicaria*, which at first I suspected might have proved your willow-lark,[1] but, on a nicer examination, it answered much better to the description of that species which you shot at Revesby,[2] in Lincolnshire. My bird I describe thus: 'It is a size less than the grasshopper-lark; the head, back, and coverts of the wings, of a dusky brown, without those dark spots of the grasshopper-lark; over each eye is a milk-white stroke; the chin and throat are white, and the under parts of a yellowish white; the rump is tawny, and the feathers of the tail sharp-pointed; the bill is dusky and sharp, and the legs are dusky; the hinder claw long and crooked.' The person that shot it says that it sung so like a reed-sparrow that he took it for one; and that it sings all night: but this account merits farther inquiry. For my part, I suspect it is a second sort of *locustela*, hinted at by Dr Derham in Ray's Letters: see p. 108. He also procured me a grasshopper-lark.

Ring-ousel

The question that you put with regard to those genera of animals that are peculiar to America, viz., how they came there, and whence? is too puzzling for me to answer; and yet so obvious as often to have struck me with wonder. If one looks into the writers on that subject little satisfaction is to be found. Ingenious men will readily advance plausible arguments to support whatever theory they shall choose to maintain; but then the misfortune is, every one's hypothesis is each as good as another's, since they are all founded on conjecture. The late writers of this sort, in whom may be seen all the arguments of those that have gone before, as I remember stock America from the western coast of Africa and the south of Europe; and then break down the Isthmus that bridged over the Atlantic. But this is making use of a violent piece of machinery; it is a difficulty worthy of the interposition of a god! '*Incredulus odi*.'

1 For this *Salicaria*, see letter, August 30th, 1769.
2 The seat of Sir Joseph Banks. ED.

To Thomas Pennant, Esquire

THE NATURALIST'S SUMMER-EVENING WALK

. . . . equidem credo, quia sit divinitus illis
Ingenium. VIRGIL'S *Georgics*

When day declining sheds a milder gleam,
What time the may-fly[3] haunts the pool or stream;
When the still owl skims round the grassy mead,
What time the timorous hare limps forth to feed;
Then be the time to steal adown the vale,
And listen to the vagrant[4] cuckoo's tale;
To hear the clamorous[5] curlew call his mate,
Or the soft quail his tender pain relate;
To see the swallow sweep the dark'ning plain
Belated, to support her infant train;
To mark the swift in rapid giddy ring
Dash round the steeple, unsubdued of wing:
Amusive birds! – say where your hid retreat
When the frost rages and the tempests beat;
Whence your return, by such nice instinct led,
When spring, soft season, lifts her bloomy head?
Such baffled searches mock man's prying pride,
The GOD of NATURE is your secret guide!
While deepening shades obscure the face of day
To yonder bench leaf-shelter'd let us stray,
'Till blended objects fail the swimming sight,
And all the fading landscape sinks in night;
To hear the drowsy dor come brushing by
With buzzing wing, or the shrill[6] cricket cry;
To see the feeding bat glance through the wood;

3 The angler's may-fly, the *ephemera vulgata*, LINNEAUS, comes forth from its aurelia state, and emerges out of the water about six in the evening, and dies about eleven at night, determining the date of its fly state in about five or six hours. They usually begin to appear about the 4th of June, and continue in succession for near a fortnight. See *Swammerdam*, *Derham*, *Scopoli*, &c.

4 Vagrant cuckoo; so called because, being tied down by no incubation or attendance about the nutrition of its young, it wanders without control.

5 *Charadrius oedicnemus.*

6 *Gryllus campstris.*

To catch the distant falling of the flood;
While o'er the cliff th' awaken'd churn-owl hung
Through the still gloom protracts his chattering song;
While high in the air, and poised upon his wings,
Unseen, the soft enamour'd[7] woodlark sings:
These, Nature's works, the curious mind employ,
Inspire a soothing melancholy joy:
As fancy warms, a pleasing kind of pain
Steals o'er the cheek, and thrills the creeping vein!
Each rural sight, each sound, each smell, combine;
The tinkling sheep-bell, or the breath of kine;
The new-mown hay that scents the swelling breeze,
Or cottage-chimney smoking through the trees.
The chilling night-dews fall: - away, retire!
For see, the glow-worm lights her amorous fire![8]
Thus, ere night's veil had half obscured the sky,
Th' impatient damsel hung her lamp on high:
True to the signal, by love's meteor led,
Leander hasten'd to his Hero's bed.[9]

I am, &c.

7 In hot summer nights woodlarks soar to a prodigious height, and hang singing in the air.
8 The light of the female glow-worm (as she often crawls up the stalk of a grass to make herself more conspicuous) is a signal to the male, which is a slender dusky scarabaeus
9 See the story of Hero and Leander.

The Wakes

Letter 25 also to Thomas Pennant, Esq.

Selborne, August 30th, 1769

Dear Sir – It gives me satisfaction to find that my account of the ousel migration pleases you. You put a very shrewd question when you ask me how I know that their autumnal migration is southward? Was not candour and openness the very life of natural history, I should pass over this query just as a sly commentator does over a crabbed passage in a classic; but common ingenuousness obliges me to confess, not without some degree of shame, that I only reasoned in that case from analogy. For as all other autumnal birds migrate from the northward to us, to partake of our milder winters, and return to the northward again when the rigorous cold abates, so I concluded that the ring-ousels did the same, as well as their congeners the fieldfares; and especially as ring-ousels are known to haunt cold mountainous countries: but I have good reason to suspect since that they may come to us from the westward; because I

Whitethroat

hear, from very good authority, that they breed on Dartmoor; and that they forsake that wild district about the time that our visitors appear, and do not return till late in the spring.

I have taken a great deal of pains about your *salicaria* and mine, with a white stroke over its eye and a tawny rump. I have surveyed it alive and dead, and have procured several specimens, and am perfectly persuaded myself (and trust you will soon become convinced of the same) that it is no more nor less than the *passer arundinaceus minor* of Ray. This bird, by some means or other, seems to be entirely omitted in the *British Zoology*; and one reason probably was because it is so strangely classed in Ray, who ranges it among his *Pici affines.* It ought no doubt to have gone among his *Aviculae cauda unicolore*, and among your slender-billed small birds of the same division. Linnaeus might with great propriety have put it into his genus of *motacilla*; and *motacilla salicaria* of *his fauna suecica* seems to come the nearest to it. It is no uncommon bird, haunting the sides of ponds and rivers where there is covert, and the reeds and sedges of moors. The country people in some places call it the sedge-bird. It sings incessantly night and day during the breeding-time, imitating the note of a sparrow, a swallow, a skylark; and has a strange hurrying manner in its song. My specimens correspond most minutely to the description of your *fen salicaria* shot near Revesby.[1] Mr Ray has given an excellent characteristic of it when he says, '*Rostrum et pedes in hac avicula multo majores sunt quam pro corporis ratione.*'[2] See letter, May 29th, 1769

Fieldfare

I have got you the egg of an *oedicnemus*, or stone-curlew, which was picked up in a

1 This is the sedge-warbler, *Acrocephalus phragmitis.* ED.

2 The bill and feet in this bird are much too large to be proportionate with its body. ED.

fallow on the naked ground: there were two; but the finder inadvertently crushed one with his foot before he saw them.

When I wrote to you last year on reptiles, I wish I had not forgot to mention the faculty that snakes have of stinking *se defendendo*. I knew a gentleman who kept a tame snake, which was in its person as sweet as any animal while in good humour and unalarmed; but as soon as a stranger, or a dog or cat, came in, it fell to hissing, and filled the room with such nauseous effluvia as rendered it hardly supportable. Thus the squnck, or stonck [skunk], of Ray's *Synop. Quadr.* is an innocuous and sweet animal; but, when pressed hard by dogs and men, it can eject such a most pestilent and fetid smell and excrement, than which nothing can be more horrible.

A gentleman sent me lately a fine specimen of the *lanius minor cinerascens cum macula in scapulis alba, Raii*;[3] which is a bird that, at the time of your publishing your two first volumes of *British Zoology*, I find you had not seen. You have described it well from Edwards's drawing.

3 The woodchat, *Lanius pomeranus*.. ED.

Snow-bunting

Letter 26 also to Thomas Pennant, Esq.

Selborne, December 8th, 1769

Dear Sir – I was much gratified by your communicative letter on your return from Scotland, where you spent some considerable time, and gave yourself good room to examine the natural curiosities of that extensive kingdom, both those of the islands, as well as those of the highlands. The usual bane of such expeditions is hurry, because men seldom allot themselves half the time they should do; but, fixing on a day for their return, post from place to place, rather as if they were on a journey that required dispatch, than as philosophers investigating the works of nature. You must have made, no doubt, many discoveries, and laid up a good fund of materials for a future edition of the *British Zoology*; and will have no reason to repent that you have bestowed so much pains on a part of Great Britain that perhaps was never so well examined before.

It has always been matter of wonder to me that fieldfares, which are so congenerous to thrushes and blackbirds, should never choose to breed in England; but that they should not think even the highlands cold and northerly, and sequestered enough, is a circumstance still more strange and wonderful. The ring-ousel, you find, stays in Scotland the whole year round; so that we have reasons to conclude that those migrators that visit us for a short space every autumn do not come from thence.

And here, I think, will be the proper place to mention that those birds were most punctual again in their migration this autumn, appearing, as before, about the 30th of September; but their flocks were larger than common, and their stay protracted somewhat beyond the usual time. If they came to spend the whole winter with us, as some of their congeners do, and then left us, as they do, in spring, I should not be so much struck with the occurrence, since it would be similar to that of the other winter birds of passage; but when I see them for a fortnight at Michaelmas, and again for about a week in the middle of April, I am seized with wonder, and long to be informed whence these travellers come, and whither they go, since they seem to use our hills merely as an inn or baiting-place.

Your account of the greater brambling, or snow-fleck, is very amusing; and strange it is that such a short-winged bird should delight in such perilous voyages over the northern ocean! Some country people in the winter-time have every now and then told me that they have seen two or three white larks on our downs; but, on considering the matter, I begin to suspect that these are some stragglers of the birds we are talking of, which sometimes perhaps may rove so far to the southward.

It pleases me to find that white hares are so frequent on the Scottish mountains, and especially as you inform me that it is a distinct species; for the quadrupeds of Britain are so few, that every new species is a great acquisition.

The eagle-owl, could it be proved to belong to us, is so majestic a bird, that it would grace our *fauna* much. I never was informed before where wild-geese are known to breed.

You admit, I find, that I have proved your *fen salicaria* to be the lesser reed-sparrow of Ray; and I think you may be secure that I am right, for I took very particular pains to clear up that matter, and had some fair specimens; but, as they were not well preserved, they are decayed already. You will, no doubt, insert it in its proper place in your next edition. Your additional plates will much improve your work.

De Buffon, I know, has described the water shrew-mouse: but still I am pleased to find you have discovered it in Lincolnshire, for the reason I have given in the article of the white hare.

As a neighbour was lately ploughing in a dry chalky field, far removed from any water, he turned out a water-rat, that was curiously laid up in an hybernaculum artificially formed of grass and leaves. At one end of the burrow lay above a gallon of potatoes regularly stowed, on which it was to have supported itself for the winter. But the difficulty with me is how this *amphibius mus* came to fix its winter station at such a distance from the water. Was it determined in its choice of that place by the mere accident of finding the potatoes which were planted there; or is it the constant practice of the aquatic rat to forsake the neighbourhood of the water in the colder months?

Though I delight very little in analogous reasoning, knowing how fallacious it is with respect to natural history; yet, in the following instance, I cannot help being inclined to think it may conduce towards the explanation of a difficulty that I have mentioned before, with respect to the invariable early retreat of the *hirundo apus*, or swift, so many weeks before its congeners; and that not only with us, but also in Andalusia, where they also begin to retire about the beginning of August.

The great large bat[1] (which by the by is at present a nondescript in England, and what I have never been able yet to procure) retires or migrates very early in the summer; it also ranges very high for its food, feeding in a different region of the air; and that is the reason I never could procure one. Now this is exactly the case with the swifts; for they take their food in a more exalted region than the other species, and are very seldom seen hawking for flies near the ground, or over the surface of the water. From hence I would conclude that these hirundines and the larger bats are supported by some sorts of high-flying gnats, scarabs, or *phalaenae*, that are of short continuance; and that the short stay of these strangers is regulated by the defect of their food.

By my journal it appears that curlews clamoured on to October the thirty-first; since which I have not seen or heard any. Swallows were observed on to November the third.

1 The little bat appears almost every month in the year; but I have never seen the large ones till the end of April, nor after July. They are most common in June, but never in any plenty: are a rare species with us.

Hedgehog

Letter 27 also to Thomas Pennant, Esq.

Selborne, February 22nd, 1770

Dear Sir – Hedgehogs abound in my gardens and fields. The manner in which they eat the roots of the plantain in my grass-walks is very curious; with their upper mandible, which is much longer than their lower, they bore under the plant, and so eat the root off upwards, leaving the tufts of leaves untouched. In this respect they are serviceable, as they destroy a very troublesome weed; but they deface the walks in some measure by digging little round holes. It appears, by the dung that they drop upon the turf, that beetles are no inconsiderable part of their food. In June last I procured a litter of four or five young hedgehogs, which appeared to be about five or six days old; they, I find, like puppies, are born blind, and could not see when they came to my hands. No doubt their spines are soft and flexible at the time of their birth, or else the poor dam would have but a bad time of it in the critical moment of parturition; but it is plain they soon harden; for these little pigs had such stiff prickles on their backs and sides as would easily have fetched blood, had they not been handled with caution. Their spines are quite white at this age; and they have little hanging ears, which I do not remember to be discernible in the old ones. They can, in part, at this age draw their skin down over their faces; but are not able to contract themselves into a ball, as they do, for the sake of defence, when full grown. The reason, I suppose, is, because the curious muscle that enables the creature to roll itself up in a ball was not then arrived at its full tone and firmness. Hedgehogs make a deep and warm *hybernaculum* with leaves and moss, in which they conceal themselves for the winter: but I never could find that they stored in any winter provision, as some quadrupeds certainly do.

I have discovered an anecdote with respect to the fieldfare (*Turdus pilaris*), which I think is particular enough; this bird, though it sits on trees in the day-time, and procures the greatest part of its food from white-thorn hedges; yea, moreover, builds on very high trees, as may be seen by the *Fauna Suecica*; yet always appears with us to roost on the ground. They are seen to come in flocks just before it is dark, and to settle and nestle among the heath on our forest. And besides, the larkers in dragging their nets by night frequently catch them in the wheat stubbles; while the bat-fowlers, who take many redwings in the hedges, never entangle any of this species. Why these birds, in the matter of roosting, should differ from all their congeners, and from themselves also with respect to their proceedings by day, is a fact for which I am by no means able to account.

Fieldfare

I have somewhat to inform you of concerning the moose-deer; but in general foreign animals fall seldom in my way; my little intelligence is confined to the narrow sphere of my own observations at home.

Letter 28 also to Thomas Pennant, Esq.

Selborne, March 1770

On Michaelmas Day 1768 I managed to get a sight of the female moose belonging to the Duke of Richmond, at Goodwood; but was greatly disappointed, when I arrived at the spot, to find that it died, after having appeared in a languishing way for some time, on the morning before. However, understanding that it was not stripped, I proceeded to examine this rare quadruped: I found it in an old greenhouse, slung under the belly and chin by ropes, and in a standing posture; but though it had been dead for so short a time, it was in so putrid a state that the stench was hardly supportable. The grand distinction between this deer, and any other species that I have ever met with, consisted in the strange length of its legs; on which it was tilted up much in the manner of the birds of the*grallae* order. I measured it, as they do an horse, and found that from the ground to the withers it was just five feet four inches; which height answers exactly to sixteen hands, a growth that few horses arrive at; but then, with his length of legs, its neck was remarkably short, no more than twelve inches; so that, by straddling with one foot forward and the other backward, it grazed on the plain ground, with the greatest difficulty, between its legs; the ears were vast and lopping, and as long as the neck; the head was about twenty inches long, and ass-like; and had such a redundancy of upper lip as I never saw before, with huge nostrils. This lip, travellers say, is esteemed a dainty dish in North America. It is very reasonable to suppose that this creature supports itself chiefly by browsing of trees and by wading after water plants; towards which way of livelihood the length of legs and great lip must contribute much. I have read somewhere that it delights in eating the *nymphae*, or water-lily. From the fore-feet to the belly behind the shoulder it measured three feet and eight inches: the length of the legs before and behind consisted a great deal in the *tibia*, which was strangely long; but, in my haste to get out of the stench, I forgot to measure that joint exactly. Its scut seemed to be about an inch long; the colour was a grizzled black; the mane about four inches long; the fore-hoofs were upright and shapely, the hind flat and splayed. The spring before it was only two years old, so that most probably it was not then come to its growth. What a vast tall beast must a full-grown stag be! I have been told some arrive at ten feet and an half! This poor creature had at first a female companion of the same species, which died the spring before. In the same garden was a young stag, or red deer, between whom and this moose it was hoped that there might have been a breed; but their inequality of height must have always been a bar to any commerce of the amorous kind. I should have been glad to have examined the teeth, tongue, lips, hoofs, &c., minutely; but the putrefaction precluded all farther curiosity. This animal, the keeper told me, seemed to enjoy itself best in the extreme frost of the former winter. In the house they showed me the horn of a male moose, which had no front antlers, but only a broad palm with some snags on the edge. The noble owner of the dead moose proposed to make a skeleton of her bones.

Please to let me hear if my female moose corresponds with that you saw; and whether you think still that the American moose and European elk are the same creature.[1]

I am, with the greatest esteem, &c.

1 The American moose is not now generally recognised as a distinct species from the European elk. ED.

Priory Farm

Letter 29 also to Thomas Pennant, Esq.

Selborne, May 12th, 1770

Dear Sir – Last month we had such a series of cold turbulent weather, such a constant succession of frost, and snow, and hail, and tempest, that the regular migration or appearance of the summer birds was much interrupted. Some did not show themselves (at least were not heard) till weeks after their usual time; as the blackcap and whitethroat; and some have not been heard yet, as the grasshopper-lark and largest willow-wren. As to the fly-catcher, I have not seen it; it is indeed one of the latest, but should appear about this time: and yet, amidst all this meteorous strife and war of the elements, two swallows discovered themselves as long ago as the eleventh of April, in frost and snow; but they withdrew quickly, and were not visible again for many days. House-martins, which are always more backward than swallows, were not observed till May came in.

Among the monogamous birds several are to be found after pairing-time, single, and of each sex; but whether this state of celibacy is matter of choice or necessity, it is not so easily discoverable. When the house-sparrows deprive my martins of their nests, as soon as I cause one to be shot, the other, be it cock or hen, presently procures a mate, and so for several times following.

I have known a dove-house infested by a pair of white owls, which made great havoc among the young pigeons: one of the owls was shot as soon as possible; but the survivor readily found a mate, and the mischief went on. After some time the new pair were both destroyed, and the annoyance ceased.

Another instance I remember of a sportsman, whose zeal for the increase of his game being greater than his humanity, after pairing time he always shot the cock bird of every couple of partridges upon his grounds; supposing that the rivalry of many males interrupted the breed: he used to say, that, though he had widowed the same hen several times, yet he found she was still provided with a fresh paramour, that did not take her away from her usual haunt.

Again; I knew a lover of setting, an old sportsman, who has often told me that soon after harvest he has frequently taken small coveys of partridges, consisting of cock birds alone; these he pleasantly used to call old bachelors.

There is a propensity belonging to common house-cats that is very remarkable; I mean their violent fondness for fish, which appears to be their most favourite food: and yet nature in this instance seems to have planted in them an appetite that, unassisted, they know not how to gratify: for of all quadrupeds cats are the least disposed towards water; and will not, when they can avoid it, deign to wet a .foot, much less to plunge into that element.

Quadrupeds that prey on fish are amphibious: such is the otter, which by nature is so well formed for diving that it makes great havoc among the inhabitants of the waters. Not supposing that we had any of those beasts in our shallow brooks, I was much pleased to see a male otter brought to me, weighing twenty-one pounds, that had been shot on the bank of our stream below the Priory, where the rivulet divides the parish of Selborne from Harteley Wood.

Raven

Letter 30 also to Thomas Pennant, Esq.

Selborne, August 1st, 1770

Dear Sir – The French, I think, in general are strangely prolix in their natural history. What Linnaeus says with respect to insects holds good in every other branch: *'Verbositas praesentis saeculi, calamitas artis.'*

Pray how do you approve of Scopoli's new work? As I admire his *Entomologia*, I long to see it.

I forgot to mention in my last letter (and had not room to insert in the former) that the male moose, in rutting-time, swims from island to island, in the lakes and rivers of North America, in pursuit of the females. My friend, the chaplain, saw one killed in the water as it was on that errand in the river St Lawrence: it was a monstrous beast, he told me; but he did not take the dimensions.

When I was last in town our friend Mr Barrington most obligingly carried me to see many curious sights. As you were then writing to him about horns, he carried me to see many strange and wonderful specimens. There is, I remember, at Lord Pembroke's, at Wilton, an horn room furnished with more than thirty different pairs; but I have not seen that house lately.

Mr Barrington showed me many astonishing collections of stuffed and living birds from all quarters of the world. After I had studied over the latter for a time, I remarked that every species almost that came from distant regions, such as South America, the coast of Guinea, &c., were thick-billed birds of the *loxia* and *fringilla* genera; and no *motacilla*, or *muscicapae*, were to be met with. When I came to consider, the reason was obvious enough; for the hard-billed birds subsist on seeds which are easily carried on board; while the softbilled birds, which are supported by worms and insects, or, what is a *succedaneum* for them, fresh raw meat, can meet with neither in long and tedious voyages. It is from this defect of food that our collections (curious as they are) are defective, and we are deprived of some of the most delicate and lively genera.

I am, &c.

At the end of the village

Letter 31 also to Thomas Pennant, Esq.

Selborne, September 14th, 1770

Dear Sir – You saw, I find, the ring-ousels again among their native crags; and are farther assured that they continue resident in those cold regions the whole year. From whence then do our ring-ousels migrate so regularly every September, and make their appearance again, as if in their return, every April? They are more early this year than common, for some were seen at the usual hill on the fourth of this month.

An observing Devonshire gentleman tells me that they frequent some parts of Dartmoor, and breed there; but leave those haunts about the end of September, or beginning of October, and return again about the end of March.

Another intelligent person assures me that they breed in great abundance all over the peak of Derby, and are called there tor-ousels; withdraw in October and November, and return in spring. This information seems to throw some light on my new migration.

Scopoli's[1] new work (which I have just procured) has its merit in ascertaining many of the birds of the Tirol and Carniola. Monographers, come from whence they may, have, I think, fair pretence to challenge some regard and approbation from the lovers of natural history; for, as no man can alone investigate all the works of nature, these partial writers may, each in their department, be more accurate in their discoveries, and freer from errors, than more general writers; and so by degrees may pave the way to an universal correct natural history. Not that Scopoli is so circumstantial and attentive to the life and conversation of his birds as I could wish: he advances some false facts; as when he says of the *hirundo urbica* that '*pullos extra nidum non nutrit*'. This assertion I know to be wrong from repeated observation this summer; for house-martins do feed their young flying, though it must be acknowledged not so commonly as the house-swallow; and the feat is done in so quick a manner as not to be perceptible to indifferent observers. He also advances some (I was going to say) improbable facts; as when he says of the woodcock that '*pullos rostro portat fugiens ab hoste*'. But candour forbids me to say absolutely that any fact is false, because I have never been witness to such a fact. I have only to remark that the long unwieldy bill of the woodcock is perhaps the worst adapted of any among the winged creation for such a feat of natural affection.

I am, &c.

1 The *Annus Primus Historico-Naturalis*. ED.

The Wishing Stone

Letter 32 also to Thomas Pennant, Esq.

Selborne, October 29th, 1770

Dear Sir – After an ineffectual search in Linnaeus, Brisson, &c., I begin to suspect that I discern my brother's *hirundo hyberna* in Scopoli's new-discovered *hirundo rupestris*, p. 167. His description of '*Supra murina, subtus albida; rectrices macula ovali allba in latere interno; pedes nudi, nigri; rostrum nigrum; remiges obscuriores quam plumae dorsales; rectrices remigibus concolores; cauda emarginata, nec forcipata;*' agrees very well with the bird in question: but when he comes to advance that it is '*statura hirundinis urbicae*', and that '*definitio hirundinis ripariae Linnaei huic quoque conveniït*', he in some measure invalidates all he has said; at least he shows at once that he compares them to these species merely from memory; for I have compared the birds themselves, and find they differ widely in every circumstance of shape, size, and colour. However, as you will have a specimen, I shall be glad to hear what your judgment is in the matter.[1]

1 This was the *Hirundo rupestris* of Linnaeus, of which John White, Gilbert White's brother, who was chaplain at Gibraltar, had sent specimens from that place to the Swedish naturalist. ED.

Whether my brother is forestalled in his nondescript or not, he will have the credit of first discovering that they spend their winters under the warm and sheltery shores of Gibraltar and Barbary.

Scopoli's characters of his *ordines* and *genera* are clear, just, and expressive, and much in the spirit of Linnaeus. These few remarks are the result of my first perusal of Scopoli's *Annus Primus.*

The bane of our science is the comparing one animal to the other by memory: for want of caution in this particular Scopoli falls into errors: he is not so full with regard to the manners of his indigenous birds as might be wished, as you justly observe: his Latin is easy, elegant, and expressive, and very superior to Kramer's.[2]

I am pleased to see that my description of the moose corresponds so well with yours.

I am, &c.

2 See his *Elenchus Vegetabilium et Animalium per Austriam Inferiorem, &c* .

Gibraltar swift

Letter 33 also to Thomas Pennant, Esq.

Selborne, November 26th, 1770

Dear Sir – was much pleased to see, among the collection of birds from Gibraltar, some of those short-winged English summer birds of passage, concerning whose departure we have made so much inquiry. Now if these birds are found in Andalusia to migrate to and from Barbary, it may easily be supposed that those that come to us may migrate back to the continent, and spend their winters in some of the warmer parts of Europe. This is certain, that many soft-billed birds that come to Gibraltar appear there only in spring and autumn, seeming to advance in pairs towards the northward, for the sake of breeding during the summer months; and retiring in parties and broods towards the south at the decline of the year: so that the rock of Gibraltar is the great rendezvous, and place of observation, from whence they take their departure each way towards Europe or Africa. It is therefore no mean discovery, I think, to find that our small short-winged summer birds of passage are to be seen spring and autumn on the very skirts of Europe; it is a presumptive proof of their emigrations.

Scopoli seems to me to have found the *Hirundo melba*, the great Gibraltar swift, in Tirol, without knowing it. For what is his *Hirundo alpina* but the afore-mentioned bird in other words? Says he '*Omnia prioris*' (meaning the swift); '*sed pectus album; paulo major priore*.' I do not suppose this to be a new species. It is true also of the *melba*, that '*nidificat in excelsis Alpium rupibus*'. *Vide Annum Primum*[1]

My Sussex friend, a man of observation and good sense, but no naturalist, to whom I applied on account of the stone-curlew, *oedicnemus*, sends me the following account: 'In looking over my *Naturalist's Journal* for the month of April, I find the stone-curlews are first mentioned on the seventeenth and eighteenth, which date seems to me rather late. They live with us all the spring and summer, and at the beginning of autumn prepare to take leave by getting together in flocks. They seem to me a bird of passage that may travel into some dry hilly country south of us, probably Spain, because of the abundance of sheep-walks in that country; for they spend their summers with us in such districts. This conjecture I hazard, as I have never met with any one that has seen them in England in the winter. I believe they are not fond of going near the water, but feed on earthworms, that are common on sheep-walks and downs. They breed on fallows and lay-fields abounding with grey mossy flints, which much resemble their young in colour; among which they skulk and conceal themselves. They make no nest, but lay their eggs on the bare ground, producing in common but two at a time. There is reason to think their young run soon after they are hatched; and that the old ones do not feed them, but only lead them about at the time of feeding, which, for the most part, is in the night.' Thus far, my friend.

In the manners of this bird you see there is something very analogous to the bustard, whom it also somewhat resembles in aspect and make, and in the structure of its feet.

For a long time I have desired my relation to look out for these birds in Andalusia; and now he writes me word that, for the first time, he saw one dead in the market on the third of September.

When the *oedicnemus* flies it stretches out its legs straight behind, like an heron.I am, &c.

1 This is the *Cypselus melba*, sent to Linnaeus by John White from Gibraltar. It is now known that swifts are not swallows, nor related to the swallow, the resemblance between the two being merely external and due to similarity of habit. ED.

Basingstoke Grammar School

Letter 34 also to Thomas Pennant, Esq.

Selborne, March 30th, 1771

Dear Sir – There is an insect with us, especially on chalky districts, which is very troublesome and teasing all the latter end of the summer, getting into people's skins, especially those of women and children, and raising tumours which itch intolerably. This animal (which we call an harvest bug) is very minute, scarce discernible to the naked eye; of a bright scarlet colour, and of the genus of Acarus. They are to be met with in gardens on kidney-beans, or any legumens, but prevail only in the hot months of summer.

Warreners, as some have assured me, are much infested by them on chalky downs; where these insects swarm sometimes to so infinite a degree as to discolour their nets, and to give them a reddish cast, while the men are so bitten as to be thrown into fevers.

There is a small long shining fly in these parts very troublesome to the housewife, by getting into the chimneys, and laying its eggs in the bacon while it is drying; these eggs produce maggots called jumpers, which, harbouring in the gammons and best parts of the hogs, eat down to the bone, and make great waste. This fly I suspect to be a variety of the *Musca putris* of Linnaeus; it is to be seen in the summer in farm-kitchens on the bacon-racks and about the mantelpieces, and on the ceilings.

The insect that infests turnips and many crops in the garden (destroying often whole fields while in their seedling leaves) is an animal that wants to be better known. The country people here call it the turnip-fly and black-dolphin; but I know it to be one of the *coleoptera*; the '*chrysomela oleracea*, *saltatoria*, *femoribus*, *posticis crassissimis*'. In very hot summers they abound to an amazing degree, and, as you walk in a field or in a garden, make a pattering like rain, by jumping on the leaves of the turnips or cabbages.

There is an oestrus, known in these parts to every ploughboy which, because it is omitted by Linnaeus, is also passed over by late writers; and that is the *curvicauda* of old Mouset, mentioned by Derham in his *Physico-Theology*, p. 250; an insect worthy of remark for depositing its eggs as it flies in so dexterous a manner on the single hairs of the legs and flanks of grass-horses. But then Derham is mistaken when he advances that this oestrus is the parent of that wonderful star-tailed maggot which he mentions afterwards; for more modern entomologists have discovered that singular production to be derived from the egg of the *Musca chameleon*; see Geoffroy, t. xvii. f. 4.[1]

A full history of noxious insects hurtful in the field, garden, and house, suggesting all the known and likely means of destroying them, would be allowed by the public to be a most useful and important work. What knowledge there is of this sort lies scattered, and wants to be collected; great improvements would soon follow of course. A knowledge of the properties, economy, propagation, and in short of the life and conversation of these animals, is a necessary step to lead us to some method of preventing their depredations.

As far as I am a judge, nothing would recommend entomology more than some neat plates that should well express the generic distinctions of insects according to Linnaeus; for I am well assured that many people would study insects, could they set out with a more adequate notion of those distinctions than can be conveyed at first by words alone.

1 White is here mistaken. ED.

Peacock

Letter 35 also to Thomas Pennant, Esq.

Selborne, 1771

Dear Sir – Happening to make a visit to my neighbour's peacocks, I could not help observing that the trains of those magnificent birds appear by no means to be their tails; those long feathers growing not from their *uropygium*, but all up their backs. A range of short brown stiff feathers, about six inches long, fixed in the *uropygium*, is the real tail, and serves as the *fulcrum* to prop the train, which is long and top-heavy when set on end. When the train is up, nothing appears of the bird before but its head and neck; but this would not be the case were those long feathers fixed only in the rump, as may be seen by the turkey-cock when in a strutting attitude. By a strong muscular vibration these birds can make the shafts of their long feathers clatter like the swords of a sword-dancer; they then trample very quick with their feet, and run backwards towards the females.

I should tell you that I have got an uncommon *calculus aegogropila*, taken out of the stomach of a fat ox; it is perfectly round, and about the size of a large Seville orange: such are, I think, usually flat.

Letter 36 also to Thomas Pennant, Esq.

September 1771

Dear Sir – The summer through I have seen but two of that large species of bat which I call *Vespertilio altivolans*, from its manner of feeding high in the air; I procured one of them, and found it to be a male; and made no doubt, as they accompanied together, that the other was a female; but, happening in an evening or two to procure the other likewise, I was somewhat disappointed, when it appeared to be also of the same sex. This circumstance, and the great scarcity of this sort, at least in these parts, occasions some suspicions in my mind whether it is really a species, or whether it may not be the male part of the more known species, one of which may supply many females; as is known to be the case in sheep and some other quadrupeds. But this doubt can only be cleared by a farther examination, and some attention to the sex, of more specimens; all that I know at present is, that my two were amply furnished with the parts of generation, much resembling those of a boar.

In the extent of their wings they measured fourteen inches and a half; and four inches and a half from the nose to the tip of the tail; their heads were large, their nostrils bilobated, their shoulders broad and muscular; and their whole bodies fleshy and plump. Nothing could be more sleek and soft than their fur, which was of a bright chestnut colour; their maws were full of food, but so macerated that the quality could not be distinguished; their livers, kidneys, and hearts, were large, and their bowels covered with fat. They weighed each, when entire, full one ounce and one drachm. Within the ear there was somewhat of a peculiar structure that I did not understand perfectly; but refer it to the observation of the curious anatomist. These creatures sent forth a very rancid and offensive smell.

Great bat

Nightjar
or Goatsucker

Letter 37 also to Thomas Pennant, Esq.

Selborne, 1771

Dear Sir – On the twelfth of July I had a fair opportunity of contemplating the motions of the *caprimulgus*, or fern-owl, as it was playing round a large oak that swarmed with *Scarabaei solstitiales*, or fern-chafers. The powers of its wing were wonderful, exceeding, if possible, the various evolutions and quick turns of the swallow genus. The circumstance that pleased me most was, that I saw it distinctly, more than once, put out its short leg while on the wing, and, by a bend of the head, deliver somewhat into its mouth. If it takes any part of its prey with its foot, as I have now the greatest reason to suppose it does these chafers, I no longer wonder at the use of its middle toe, which is curiously furnished with a serrated claw.

Swallows and martins, the bulk of them I mean, have forsaken us sooner this year than usual; for on September the twenty-second they rendezvoused in a neighbour's walnut-tree, where it seemed probable they had taken up their lodging for the night. At the dawn of the day, which was foggy, they arose all together in infinite numbers, occasioning such a rushing from the strokes of their wings against the hazy air, as might be heard to a considerable distance: since that no flock has appeared, only a few stragglers.

Some swifts stayed late, till the twenty-second of August – a rare instance! for they usually withdraw within the first week.[1]

On September the twenty-fourth three or four ring-ousels appeared in my fields for the first time this season; how punctual are these visitors in their autumnal and spring migrations!

1 See Letter 53 to Mr Barrington.

Old Hop Kilns.

Letter 38 also to Thomas Pennant, Esq.

Selborne, March 15th, 1773

Dear Sir – By my journal for last autumn it appears that the house-martins bred very late, and stayed very late in these parts; for, on the first of October, I saw young martins in their nest nearly fledged; and again on the twenty-first of October, we had at the next house a nest full of young martins just ready to fly; and the old ones were hawking insects with great alertness. The next morning the brood forsook their nest, and were flying round the village. From this day I never saw one of the swallow kind till November the third; when twenty, or perhaps thirty, house-martins were playing all day long by the side of the hanging wood, and over my field. Did these small weak birds, some of which were nestling twelve days ago, shift their quarters at this late season of the year to the other side of the northern tropic? Or rather, is it not more probable that the next church, ruin, chalk-cliff, steep covert, or perhaps sandbank, lake or pool (as a more northern naturalist would say), may become their *hybernaculum*, and afford them a ready and obvious retreat?

We now begin to expect our vernal migration of ring-ousels every week. Persons worthy of credit assure me that ring-ousels were seen at Christmas 1770 in the forest of Bere, on the southern verge of this county. Hence we may conclude that their migrations are only internal, and not extended to the continent southward, if they do at first come at all from the northern parts of this island only, and not from the north of Europe. Come from whence they will, it is plain, from the fearless disregard that they show for men or guns, that they have been little accustomed to places of much resort. Navigators mention that in the Isle of Ascension, and other such desolate districts, birds are so little acquainted with the human form that they settle on men's shoulders; and have no more dread of a sailor than they would have of a goat that was grazing. A young man at Lewes, in Sussex, assured me that about seven years ago ring-ousels abounded so about that town in the autumn that he killed sixteen himself in one afternoon; he added further, that some had appeared since in every autumn; but he could not find that any had been observed before the season in which he shot so many. I myself have found these birds in little parties in the autumn cantoned all along the Sussex downs, wherever there were shrubs and bushes, from Chichester to Lewes; particularly in the autumn of 1770.

I am, &c.

Osprey

Letter 39 also to Thomas Pennant, Esq.[1]

Selborne, November 9th, 1773

Dear Sir – As you desire me to send you such observations as may occur, I take the liberty of making the following remarks, that you may, according as you think me right or wrong, admit or reject what I here advance, in your intended new edition of the *British Zoology*.

The osprey[2] was shot about a year ago at Frinsham [Frensham] Pond, a great lake, at about six miles from hence, while it was sitting on the handle of a plough and devouring a fish: it used to precipitate itself into the water, and so take its prey by surprise.

A great ash-coloured[3] butcher-bird was shot last winter in Tisted

1 *British Zoology*, vol. i. p. 128.

2 This and the following letter were evidently written at Pennant's request as material for his *British Zoology*, and were used by him on the various pages referred to below. ED.

3 *Op. cit.*, vol. i. p. 161.

Park, and a red-backed butcher-bird at Selborne: they are *rarae aves* in this county.

Crows[4] go in pairs all the year round.

Cornish choughs[5] abound, and breed on Beechy Head, and on all the cliffs of the Sussex coast.

The common wild-pigeon,[6] or stock-dove, is a bird of passage in the south of England, seldom appearing till towards the end of November; is usually the latest winter-bird of passage. Before our beechen woods were so much destroyed, we had myriads of them, reaching in strings for mile a together as they went out in a morning to feed. They leave us early in spring: where do they breed?

The people of Hampshire and Sussex call the missel-bird[7] the storm-cock, because it sings early in the spring in blowing showery weather; its song often commences with the year: with us it builds much in orchards.

A gentleman assures me he has taken the nests of ring-ousels[8] on Dartmoor: they build in banks on the sides of streams.

Titlarks[9] not only sing sweetly as they sit on trees, but also as they play and toy about on the wing; and particularly while they are descending, and sometimes as they stand on the ground.[10]

Adanson's[11] testimony seems to me to be a very poor evidence that European swallows migrate during our winter to Senegal: he does not talk at all like an ornithologist; and probably saw only the swallows of that country, which I know build within Governor O'Hara's hall against the roof. Had he known European swallows, would he not have mentioned the species?

The house-swallow washes by dropping into the water as it flies: this species appears commonly about a week before the house-martin, and about ten or twelve days before the swift.

In 1772 there were young house-martins[12] in their nest till October the twenty-third.

4 *Ibid.*, Vol. i. p. 167.
5 *Ibid.*, Vol. i. p. 198.
6 *Ibid.*, Vol. i. p. 216.
7 *Ibid.*, Vol. i. p. 224.
8 *Ibid.*, Vol. i. p. 229.
9 *Ibid.*, Vol. ii. p. 237.
10 This is true of the tree-pipit, *Anthus trivialis*, not of the common titlark or meadow-pipit, *A. pratensis*, two birds which White apparently confuses. ED.
11 *Ibid.*, Vol. ii. p. 242.
12 *Op. cit.*, Vol. ii. p. 244.

The swift[13] appears about ten or twelve days later than the house-swallow: viz., about the twenty-fourth or twenty-sixth of April.

Whin-chats and stone-chatters[14] stay with us the whole year.

Some wheat-ears[15] continue with us the winter through.

Wagtails, all sorts, remain with us all the winter.

Bullfinches,[16] when fed on hemp-seed, often become wholly black.

We have vast flocks of female chaffinches[17] all the winter, with hardly any males among them.

When you say that in breeding-time the cock snipe[18] make a bleating noise, and I a drumming (perhaps I should have rather said an humming), I suspect we mean the same thing. However, while they are playing about on the wing they certainly make a loud piping with their mouths: but whether that bleating or humming is ventriloquous, or proceeds from the motion of their wings, I cannot say; but this I know, that when this noise happens the bird is always descending, and his wings are violently agitated.

Soon after the lapwings[19] have done breeding they congregate, and, leaving the moors and marshes, betake themselves to downs and sheep-walks.

Two years ago[20] last spring the little auk was found alive and unhurt, but fluttering and unable to rise, in a lane a few miles from Alresford, where there is a great lake: it was kept a while, but died.

I saw young teals[21] taken alive in the ponds of Wolmer Forest in the beginning of July last, along with flappers, or young wild ducks.

Speaking of the swift,[22] that page says 'its drink the dew'; whereas it should be 'it drinks on the wing'; for all the swallow kind sip their water as they sweep over the face of pools or rivers: like Virgil's bees, they drink flying; *flumina summa libant*. In this method of drinking perhaps this genus may be peculiar.

Of the sedge-bird[23] be pleased to say it sings most part of the night; its notes are hurrying, but not unpleasing, and imitative of several birds; as the sparrow, swallow, skylark. When it happens to be silent in the night, by throwing a stone or clod into the bushes

13 *Ibid.*, Vol. ii. p. 245.
14 *Ibid.*, Vol. ii. pp. 270, 271.
15 *Ibid.*, Vol. ii. p. 269.
16 *Ibid.*, Vol. ii. p. 300.
17 *Ibid.*, Vol. ii. p. 306.
18 *Ibid.*, Vol. ii. p. 358.
19 *Ibid.*, Vol. ii. p. 360.
20 *Ibid.*, Vol. ii. p. 409.
21 *Ibid.*, Vol. ii. p. 475.
22 *Ibid.*, Vol. ii. p. 15.
23 *Ibid.*, Vol. ii. p. 16.

where it sits you immediately set it a-singing; or in other words, though it slumbers sometimes, yet as soon as it is awakened it resumes its song.

White's Sundial

Nightingale

Letter 40 also to Thomas Pennant, Esq.

Selborne, September 2nd, 1774

Dear Sir – Before your letter arrived, and of my own accord, I had been remarking and comparing the tails of the male and female swallow, and this ere any young broods appeared; so that there was no danger of confounding the dams with their *pulli*: and besides, as they were then always in pairs, and busied in the employ of nidification, there could be no room for mistaking the sexes, nor the individuals of different chimneys the one for the other. From all my observations, it constantly appeared that each sex has the long feathers in its tail that give it that forked shape; with this difference, that they are longer in the tail of the male than in that of the female.

Nightingales, when their young first come abroad, and are helpless, make a plaintive and a jarring noise; and also a snapping or cracking, pursuing people along the hedges as they walk: these last sounds seem intended for menace and defiance.

The grasshopper-lark chirps all night in the height of summer.

Swans turn white the second year, and breed the third.

Weasels prey on moles, as appears by their being sometimes caught in mole-traps.

Sparrow-hawks sometimes breed in old crows' nests, and the kestrel in churches and ruins.

There are supposed to be two sorts of eels in the island of Ely. The threads sometimes discovered in eels are perhaps their young: the generation of eels is very dark and mysterious.[1]

Hen-harriers breed on the ground, and seem never to settle on trees.

When redstarts shake their tails they move them horizontally, as dogs do when they fawn: the tail of a wagtail, when in motion, bobs up and down like that of a jaded horse.

Hedge-sparrows have a remarkable flirt with their wings in breeding-time; as soon as frosty mornings come they make a very piping plaintive noise.

Many birds which become silent about Midsummer reassume their notes again in September; as the thrush, blackbird, woodlark, willow-wren, &c.; hence August is by much the most mute month, the spring, summer, and autumn through. Are birds induced to sing again because the temperament of autumn resembles that of spring?

Linnaeus ranges plants geographically; palms inhabit the tropics, grasses the temperate zones, and mosses and lichens the polar circles; no doubt animals may be classed in the same manner with propriety.

House-sparrows build under eaves in the spring; as the weather becomes hotter they get out for coolness, and nest in plum-trees and apple-trees. These birds have been known sometimes to build in rooks' nests, and sometimes in the forks of boughs under rooks' nests.

As my neighbour was housing a rick he observed that his dogs devoured all the little red mice that they could catch, but rejected the common mice; and that his cats ate the common mice, refusing the red.

Red-breasts sing all through the spring, summer, and autumn. The reason that they are called autumn songsters is, because in the two first seasons their voices are drowned and lost in the general chorus; in the latter their song becomes distinguishable. Many songsters of the autumn seem to be the young cock red-breasts of that year: notwithstanding the prejudices in their favour, they do much mischief in gardens to the summer fruits.[2]

1 The threads mentioned by White are intestinal worms. The reproduction of eels has only recently been satisfactorily understood. They never spawn in rivers, but deposit their eggs and hatch out the young in the deep sea. ED.

2 They eat also the berries of the ivy, the honeysuckle, and the *Euonymus europaes*, or spindle-tree.

The titmouse, which early in February begins to make two quaint notes, like the whetting of a saw, is the marsh titmouse: the great titmouse sings with three cheerful joyous notes, and begins about the same time.

Wrens sing all the winter through, frost excepted.

House-martins came remarkably late this year both in Hampshire and Devonshire: is this circumstance for or against either hiding or migration?

Most birds drink sipping at intervals; but pigeons take a long continued draught, like quadrupeds.

Notwithstanding what I have said in a former letter, no grey crows were ever known to breed on Dartmoor; it was my mistake.

The appearance and flying of the *Scarabaeus solstitialis*, or fernchafer, commence with the month of July, and cease about the end of it. These scarabs are the constant food of *Caprimulgi* or fernowls, through that period. They abound on the chalky downs and in some sandy districts, but not in the clays.

In the garden of the Black Bear inn in the town of Reading is a stream or canal running under the stables and out into the fields on the other side of the road: in this water are many carps, which lie rolling about in sight, being fed by travellers, who amuse themselves by tossing them bread; but as soon as the weather grows at all severe these fishes are no longer seen, because they retire under the stables, where they remain till the return of spring. Do they lie in a torpid state? if they do not, how are they supported?

The note of the white-throat, which is continually repeated, and often attended with odd gesticulations on the wing, is harsh and displeasing. These birds seem of a pugnacious disposition; for they sing with an erected crest and attitudes of rivalry and defiance; are shy and wild in breeding-time, avoiding neighbourhoods, and haunting lonely lanes and commons; nay even the very tops of the Sussex downs, where there are bushes and covert; but in July and August they bring their broods into gardens and orchards, and make great havoc among the summer-fruits.

The black-cap has in common a full, sweet, deep, loud, and wild pipe; yet that strain is of short continuance, and his motions are desultory; but when that bird sits calmly and engages in song in earnest, he pours forth very sweet, but inward melody, and expresses great variety of soft and gentle modulations, superior perhaps to those of any of our warblers, the nightingale excepted.

Black-caps mostly haunt orchards and gardens; while they warble their throats are wonderfully distended.

The song of the redstart is superior, though somewhat like that of the whitethroat; some birds have a few more notes than others. Sitting very placidly on the top of a tall tree in a village, the cock sings from morning to night: he affects neighbourhoods, and avoids solitude, and loves to build in orchards and about houses; with us he perches on the vane of a tall maypole.

The fly-catcher is of all our summer birds the most mute and the most familiar; it also appears the last of any. It builds in a vine, or a sweetbrier, against the wall of a house, or in the hole of a wall, or on the end of a beam or plate, and often close to the post of a door where people are going in and out all day long. This bird does not make the least pretension to song, but uses a little inward wailing note when it thinks its young in danger from cats or other annoyances; it breeds but once, and retires early.

Selborne parish alone can and has exhibited at times more than half the birds that are ever seen in all Sweden; the former has produced more than one hundred and twenty species, the latter only two hundred and twenty-one. Let me add also that it has shown near half the species that were ever known in Great Britain[1]

On a retrospect, I observe that my long letter carries with it a quaint and magisterial air, and is very sententious; but when I recollect that you requested stricture and anecdote, I hope you will pardon the didactic manner for the sake of the information it may happen to contain.

1 Sweden 221, Great Britain 252 species. [Many more are now known. ED.]

Great Titmouse

Letter 41 also to Thomas Pennant, Esq.

It is matter of curious inquiry to trace out how those species of soft-billed birds that continue with us the winter through, subsist during the dead months. The imbecility [feebleness] of birds seems not to be the only reason why they shun the rigour of our winters; for the robust wryneck (so much resembling the hardy race of woodpeckers) migrates, while the feeble little golden-crowned wren, that shadow of a bird, braves our severest frosts without availing himself of houses and villages, to which most of our winter birds crowd in distressful seasons, while this keeps aloof in fields and woods; but perhaps this may be the reason why they may often perish, and why they are almost as rare as any bird we know.

I have no reason to doubt but that the soft-billed birds, which winter with us, subsist chiefly on insects in their aurelia state. All the species of wagtails in severe weather haunt shallow streams near their spring-heads, where they never freeze; and, by wading, pick out the aurelias of the genus of *Phryganeae*,[1] &c.

Hedge-sparrows frequent sinks and gutters in hard weather, where they pick up crumbs and other sweepings: and in mild weather they procure worms which are stirring every month in the year, as any one may see that will only be at the trouble of taking a candle to a grass-plot on any mild winter's night. Red-breasts and wrens in the winter haunt out-houses, stables, and barns, where

1 See Derham's *Physico-theology*, p. 235.

they find spiders and flies that have laid themselves up during the cold season. But the grand support of the soft-billed birds in winter is that infinite profusion of aureliae of the *Lepidoptera ordo*, which is fastened to the twigs of trees and their trunks; to the pales and walls of gardens and buildings; and is found in every cranny and cleft of rock or rubbish, and even in the ground itself.

Every species of titmouse winters with us; they have what I call a kind of intermediate bill between the hard and the soft, between the Linnaean genera of *Fringilla* and *Motacilla*. One species alone spends its whole time in the woods and fields, never retreating for succour in the severest seasons to houses and neighbourhoods; and that is the delicate long-tailed titmouse, which is almost as minute as the golden-crowned wren; but the blue titmouse or nun (*Parus caeruleus*), the cole-mouse (*Parus ater*), the great black-headed titmouse (*Fringillago*)[2] and the marsh titmouse (*Parus palustris*), all resort at times to buildings, and in hard weather particularly. The great titmouse, driven by stress of weather, much frequents houses; and, in deep snows, I have seen this bird, while it hung with its back downwards (to my no small delight and admiration), draw straws lengthwise from out the eaves of thatched houses, in order to pull out the flies that were concealed between them, and that in such numbers that they quite defaced the thatch, and gave it a ragged appearance.

The blue titmouse, or nun, is a great frequenter of houses, and a general devourer. Besides insects, it is very fond of flesh; for it frequently picks bones on dunghills: it is a vast admirer of suet, and haunts butchers' shops. When a boy, I have known twenty in a morning caught with snap mouse-traps, baited with tallow or suet.

It will also pick holes in apples left on the ground, and be well entertained with the seeds on the head of a sunflower. The blue, marsh, and great titmice will, in very severe weather, carry away barley and oat-straws from the sides of ricks.

How the wheat-ear and whin-chat support themselves in winter cannot be so easily ascertained, since they spend their time on wild heaths and warrens; the former especially, where there are stone quarries: most probable it is that their maintenance arises from the aureliae of the *Lepidoptera ordo*, which furnish them with a plentiful table in the wilderness. I am, &c.

2 This bird is no doubt the *Parus major* of Linnaeus, the great tit or black-headed tit of most British authors. ED.

Letter 42 also to Thomas Pennant, Esq.[1]

Selborne, March 9th, 1774

Dear Sir – Some future faunist, a man of fortune, will, I hope, extend his visits to the kingdom of Ireland; a new field and a country little known to the naturalist. He will not, it is to be wished, undertake that tour unaccompanied by a botanist, because the mountains have scarcely been sufficiently examined; and the southerly counties of so mild an island may possibly afford some plants little to be expected within the British dominions. A person of a thinking turn of mind will draw many just remarks from the modern improvements of that country, both in arts and agriculture, where premiums obtained long before they were heard of with us. The manners of the wild natives, their superstitions, their prejudices, their sordid way of life, will extort from him many useful reflections. He should also take with him an able draughtsman; for he must by

1 This letter is interesting as showing the comparatively limited range of ornithologists hardly more than a century ago. Ireland was then a scarcely known country. At the present day every nook of it has been explored, zoologically and botanically, and the stations of every rare species of plant or animal exactly recorded. White was quite right in his expectation that the southern counties would afford some plants little to be expected within the United Kingdom; the flora of Kerry and Connemara abounds in essentially Spanish and Portuguese types. Nothing is more interesting in reading White than to observe the extraordinary difference in the estimate of remoteness which has been brought about by increased means of locomotion. He speaks of Ireland almost as we should now speak of New Guinea. In other letters a similar point of view may be noted with regard to Andalusia and Carniola, places now well within the ordinary tourist beat; but the expressions here used about Ireland are even more striking. The allusion to the 'lofty stupendous mountains' also gives an interesting glimpse of the eighteenth-century way of looking at nature. White's contemporaries had a marvellous faculty for standing awestruck before 'majestic heights' up which the present generation strolls easily for a picnic party. The observations on the maps of Scotland belong in the same way to what now seems a remote antiquity. ED.

no means pass over the noble castles and seats, the extensive and picturesque lakes and waterfalls, and the lofty stupendous mountains, so little known, and so engaging to the imagination when described and exhibited in a lively manner; such a work would be well received.

Blue Titmouse

As I have seen no modern map of Scotland, I cannot pretend to say how accurate or particular any such may be; but this I know, that the best old maps of that kingdom are very defective.

The great obvious defect that I have remarked in all maps of Scotland that have fallen in my way is, a want of a coloured line, or stroke, that shall exactly define the just limits of that district called the Highlands. Moreover, all the great avenues to that mountainous and romantic country want to be well distinguished. The military roads formed by General Wade are so great and Roman-like an undertaking that they well merit attention. My old map, Moll's Map, takes notice of Fort William, but could not mention the other forts that have been erected long since; therefore a good representation of the chain of forts should not be omitted.

The celebrated zigzag up the Coryarich must not be passed over. Moll takes notice of Hamilton and Drumlanrig, and such capital houses; but a new survey, no doubt, should represent every seat and castle remarkable for any great event, or celebrated for its paintings, &c. Lord Breadalbane's seat and beautiful *policy* are too curious and extraordinary to be omitted.

The seat of the Earl of Eglintoun, near Glasgow, is worthy of notice. The pine plantations of that nobleman are very grand and extensive indeed.   I am, &c.

Honey-buzzard

Letter 43 also to Thomas Pennant, Esq.

A pair of honey-buzzards, *Buteo apivorus, Linn., sive vespivorus Raii,*[1] built them a large shallow nest, composed of twigs and lined with dead beechen leaves, upon a tall slender beech near the middle of Selborne Hanger, in the summer of 1780. In the middle of the month of June a bold boy climbed this tree, though standing on so steep and dizzy a situation, and brought down an egg, the only one in the nest, which had been sat on for some time, and contained the embryo of a young bird. The egg was smaller, and not so round as those of the common buzzard; was dotted at each end with small red spots, and surrounded in the middle with a broad bloody zone.

The hen-bird was shot, and answered exactly to Mr Ray's description of that species; had a black cere, short thick legs, and a long tail. When on the wing this species may be easily distinguished from the common buzzard by its hawk-like appearance, small head, wings not so blunt, and longer tail. This specimen contained in its craw some limbs of frogs and many grey snails [slugs] without shells. The irides of the eyes of this bird were of a beautiful bright yellow colour.

1 The honey-buzzard is a very rare British bird. White is almost the only authority for any but an insect diet on its part. Most of the species undoubtedly live on the grubs and pupae of wasps and bees. ED.

About the tenth of July in the same summer a pair of sparrow hawks bred in an old crow's nest on a low beech in the same hanger; and as their brood, which was numerous, began to grow up, became so daring and ravenous, that they were a terror to all the dames in the village that had chickens or ducklings under their care. A boy climbed the tree, and found the young so fledged that they all escaped from him; but discovered that a good house had been kept: the larder was well stored with provisions; for he brought down a young blackbird, jay, and house-martin, all clean picked, and some half devoured. The old birds had been observed to make sad havoc for some days among the new-flown swallows and martins, which, being but lately out of their nests, had not acquired those powers and command of wing that enable them, when more mature, to set such enemies at defiance.

Rock-pigeon

Letter 44 also to Thomas Pennant, Esq.

Selborne, November 30th, 1780

Dear Sir – Every incident that occasions a renewal of our correspondence will ever be pleasing and agreeable to me.

As to the wild wood-pigeon, the *Oenas, or Vinago*, of Ray,[1] I am much of your mind; and see no reason for making it the origin of the common house-dove: but suppose those that have advanced that opinion may have been misled by another appellation, often given to the *Oenas*, which is that of stock-dove.

Unless the stock-dove in the winter varies greatly in manners from itself in summer, no species seems more unlikely to be domesticated, and to make an house-dove. We very rarely see the latter settle on trees at all, nor does it ever haunt the woods: but the former as long as it stays with us, from November perhaps to February, lives the same wild life with the ring-dove, *Palumbus torquatus*; frequents coppices and groves, supports itself chiefly by mast, and delights to roost in the tallest beeches. Could it be known in what manner stock-doves build, the doubt would be settled with

1 The whole question of the relation of domesticated pigeons to the wild stocks has been thoroughly investigated by Darwin, to whose classical researches the reader must be referred for more modern information. ED.

me at once, provided they construct their nests on trees, like the ring-dove, as I much suspect they do.

You received, you say, last spring a stock-dove from Sussex; and are informed that they sometimes breed in that country. But why did not your correspondent determine the place of its nidification, whether on rocks, cliffs, or trees? If he was not an adroit ornithologist I should doubt the fact, because people with us perpetually confound the stock-dove with the ring-dove.

For my own part, I readily concur with you in supposing that house doves are derived from the small blue rock-pigeon, for many reasons. In the first place the wild stock-dove is manifestly larger than the common house-dove, against the usual rule of domestication, which generally enlarges the breed. Again, those two remarkable black spots on the remiges of each wing of the stock-dove, which are so characteristic of the species, would not, one should think, be totally lost by its being reclaimed; but would often break out among its descendants. But what is worth an hundred arguments is, the instance you give in Sir Roger Mostyn's house-doves in Caernarvonshire; which, though tempted by plenty of food and gentle treatment, can never be prevailed on to inhabit their cote for any time; but, as soon as they begin to breed, betake themselves to the fastnesses of Ormshead [Great Orme's Head], and deposit their young in safety amidst the inaccessible caverns and precipices of that stupendous promontory.[2]

Naturam expellas furca . . . tamen usque recurrent.

I have consulted a sportsman, now in his seventy-eighth year, who tells me that fifty or sixty years back, when the beechen woods were much more extensive than at present, the number of wood-pigeons was astonishing; that he has often killed near twenty in a day: and that with a long wild-fowl piece he has shot seven or eight at a time on the wing as they came wheeling over his head: he moreover adds, which I was not aware of, that often there were among them little parties of small blue doves, which he calls rockiers. The food of these numberless emigrants was beech-mast and some acorns; and particularly barley, which they collected in the stubbles. But of late years, since the vast increase of turnips, that vegetable has furnished

2 The 'stupendous promontory' of the Great Orme's Head is another excellent example of the eighteenth-century point of view of nature. It is now overrun by visitors from Llandudno, and was at no time a particularly formidable eminence, except to sailors. ED.

a great part of their support in hard weather; and the holes they pick in these roots greatly damage the crop. From this food their flesh has contracted a rancidness which occasions them to be rejected by nicer judges of eating, who thought them before a delicate dish. They were shot not only as they were feeding in the fields, and especially in snowy weather, but also at the close of the evening, by men who lay in ambush among the woods and groves to kill them as they came in to roost.[3] These are the principal circumstances relating to this wonderful internal migration, which with us takes place towards the end of November, and ceases early in the spring. Last winter we had in Selborne high wood about an hundred of these doves; but in former times the flocks were so vast, not only with us but all the district round, that on mornings and evenings they traversed the air, like rooks, in strings, reaching for a mile together. When they thus rendezvoused here by thousands, if they happened to be suddenly roused from their roost-trees on an evening,

> Their rising all at once was like the sound
> Of thunder heard remote.

It will by no means be foreign to the present purpose to add, that I had a relation in this neighbourhood who made it a practice, for a time, whenever he could procure the eggs of a ring-dove, to place them under a pair of doves that were sitting in his own pigeon-house; hoping thereby, if he could bring about a coalition, to enlarge his breed, and teach his own doves to beat out into the woods and to support themselves by mast: the plan was plausible, but something always interrupted the success; for though the birds were usually hatched, and sometimes grew to half their size, yet none ever arrived at maturity. I myself have seen these foundlings in their nest displaying a strange ferocity of nature, so as scarcely to bear to be looked at, and snapping with their bills by way of menace. In short, they always died, perhaps for want of proper sustenance: but the owner thought that by their fierce and wild demeanour they frighted their foster-mothers, and so were starved.

Virgil, as a familiar occurrence, by way of simile, describes a dove haunting the cavern of a rock in such engaging numbers, that I cannot refrain from quoting the passage: and John Dryden has

3 Some old sportsmen say that the main part of these flocks used to withdraw as soon as the heavy Christmas frosts were over.

rendered it so happily in our language, that without further excuse I shall add his translation also:

Qualis spelunca subito commota Columba,
Cui domus, et dulces latebroso in pumice nidi,
Fertur in arva volans, plausumque exterrita pennis
Dat tecto ingentem - mox aere lapsa quieto,
Radit iter liquidum, celeresneque commovet alas.

As when a dove her rocky hold forsakes,
Rous'd in a fright her sounding wings she shakes;
The cavern rings with clattering: - out she flies,
And leaves her callow care, and cleaves the skies;
At first she flutters: - but at length she springs
To smoother flight, and shoots upon her wings.[4]

I am, &c.

4 This is the last letter to Pennant, and probably one written after publication of the series had been fully decided upon. It is obviously artificial. The curious habit of formally quoting Latin verses in private letters, and giving English translations of them even to readers equally well acquainted with the original, is so common, however, in eighteenth-century writers, that White may, perhaps, really have written to Pennant in this quaint fashion. A letter was in those days regarded as a serious piece of literary work, to be embellished with a neat patchwork of classical quotation. ED.

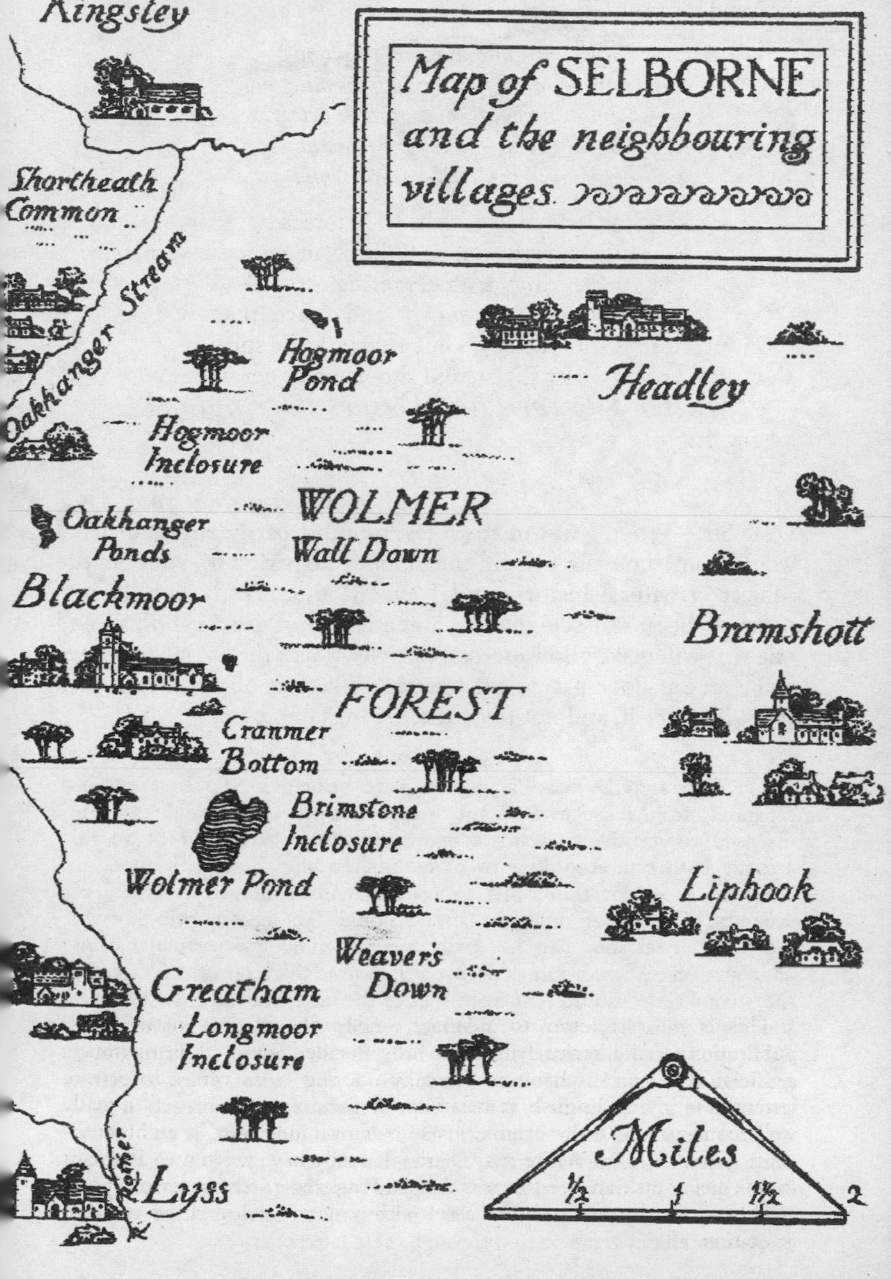

Map of SELBORNE and the neighbouring villages.

Kingsley
Shortheath Common
Oakhanger Stream
Hogmoor Pond
Headley
Hogmoor Inclosure
WOLMER
Oakhanger Ponds
Wall Down
Blackmoor
Bramshott
FOREST
Cranmer Bottom
Brimstone Inclosure
Wolmer Pond
Liphook
Weaver's Down
Greatham
Longmoor Inclosure
Miles
½ 1 1½ 2
Romer
Lyss

Blackcap

Letter 1 to the Hon. Daines Barrington[1]

Selborne, June 30th, 1769

Dear Sir – When I was in town last month I partly engaged that I would sometime do myself the honour to write to you on the subject of natural history; and I am the more ready to fulfil my promise, because I see you are a gentleman of great candour, and one that will make allowances; especially where the writer professes to be an out-door naturalist, one that takes his observations from the subject itself, and not from the writings of others.

1 The letters to Daines Barrington were printed in the first edition separately from those to Pennant, being arranged as a second part and disposed consecutively. Many subsequent editors have seen fit to re-arrange both sets according to dates, interlarding these with those to Pennant. I do not think this procedure tends either to clearness or accuracy. The reader does not always notice the superscription of the individual letter, nor can he easily bear in mind the particular 'you' addressed on each occasion. Moreover, the whole character of the letters in the second series is different from that of the letters in the first. Pennant was a naturalist who wrote to White mainly for practical information: Barrington was a dilettante theorist who generally desired confirmation of his often hasty and sometimes inaccurate *a priori* ideas. It is, therefore, undesirable to mix up the two sets, both because of the difference of their original scope, and also because, in White's own judgment, it was best to keep the personalities separate. Daines Barrington (1727–1800) was the fourth son of the first Lord Barrington, and was a barrister by profession. A dabbler – in many directions, he was a person of importance in his own day but is now chiefly remembered through these letters. ED.

THE FOLLOWING IS A LIST OF SUMMER BIRDS OF PASSAGE WHICH I HAVE DISCOVERED IN THIS NEIGHBOURHOOD, RANGED SOMEWHAT IN THE ORDER WHICH THEY APPEAR[2]

		RAII NOMINA	USUALLY APPEARS ABOUT
1	Wryneck	*Jynx, sive Torquilla*	Middle of March: harsh note
2	Smallest willow wren	*Regulus non cristatus*	March 23: chirps till September
3	Swallow	*Hirundo domestica*	April 13
4	Martin	*Hirundo rustica*	Ditto
5	Sand-martin	*Hirundo riparia*	Ditto
6	Black-cap	*Atricapilla*	Ditto: a sweet wild note
7	Nightingale	*Luscinia*	Beginning of April
8	Cuckoo	*Cuculus*	Middle of April
9	Middle willow wren	*Regulus non cristatus*	Ditto: a sweet plaintive note
10	White-throat	*Ficedulae affinis*	Ditto: mean note; sings on till September
11	Red-start	*Ruticilla*	Ditto: more agreeable song
12	Stone-curlew	*Oedicnemus*	End of March: loud nocturnal whistle
13	Turtle-dove	*Turtur*	
14	Grasshopper-lark	*Alauda minima locustae voce*	Middle April: a small sibilous note, till the end of July
15	Swift	*Hirundo apus*	April 27
I6	Less reed-sparrow	*Passer acundinaceus minor*	A sweet polyglot, but hurrying: it has the notes of many birds
17	Land-rail	*Ortygometra*	A loud harsh note, crex, crex
18	Largest willow wren	*Regulus non cristatus*	*Cantat voce stridula locustae*; end of April, on the tops of high beeches
19	Goatsucker, or fern-owl	*Caprimulgus*	Beginning of May: chatters by night with a singular noise
20	Fly-catcher	*Stoparola*	May 12: a very mute bird; this is the latest summer bird of passage

2 In this list I have not attempted to give the accepted modern scientific names. In most cases the English name sufficiently designates the birds

This assemblage of curious and amusing birds belongs to ten several genera of the Linnaean system: and are all of the *ordo* of *passeres* save the *Jynx* and *Cuculus*, which are *picae*, and the *Charadrius* (*Oedicnemus*) and *Rallus* (*Ortygometra*), which are *grallae*.

These birds, as they stand numerically, belong to the following Linnaean genera:

1	*Jynx*	13	*Colomba*
2, 6, 7, 9, 10, 11, 16, 18	*Motacilla*	17	*Rallus*
3, 4, 5, 15	*Hirundo*	19	*Caprimulgus*
8	*Cuculus*	14	*Alauda*
12	*Charadrius*	20	*Muscicapa*

Most soft-billed birds live on insects, and not on grain and seeds; and therefore at the end of summer they retire: but the following soft-billed birds, though insect eaters, stay with us the year round:

	RAII NOMINA	
Red-breast Wren	*Rubecula* *Passer troglodytei.*	These frequent houses: and haunt out-buildings in the winter: eat spiders
Hedge-sparrow	*Curruca*	Haunt sinks for crumbs and other sweepings
White-wagtail Yellow-wagtail Grey-wagtail	*Motacilla alba* *Motacilla flava* *Motacilla cinerea*	These frequent shallow rivulets near the spring heads, where they never freeze: eat the aureliae of *Phryganea*. The smallest birds that walk.
Wheat-ear	*Oenanthe*	Some of these are to be seen with us the winter through
Whin-chat	*Oenanthe secunda*	—
Stone-chatter	*Oenanthe tertia*	—
Golden-crowned wren	*Regulus cristatus*	This is the smallest British bird: haunts the top of tall trees: stays the winter through

---

intended for all who wish to identify them. Where there is doubt, as in the case of the so-called 'wild goose', or the largest willow-wren, it is not easy to decide which is the exact species that White intended. Moreover, the question is purely otiose. The nomenclature of ornithology is a very difficult subject, and no two writers are quite agreed as to the identification of early descriptions. ED.

A LIST OF THE WINTER BIRDS OF PASSAGE ROUND THIS NEIGHBOURHOOD RANGED SOMEWHAT IN THE ORDER IN WHICH THEY APPEAR

		raii nomina	
1	Ring-ousel	*Merula torquata*	This is a new migration, which I have lately discovered about Michaelmas week, and again about the 14th of March
2	Redwing	*Turdus iliacus*	About old Michaelmas
3	Fieldfare	*Turdus pilaris*	Though a percher by day, roosts on the ground
4	Royston-crow	*Cornix cinerea*	Most frequent on downs
5	Woodcock	*Scolopax*	Appears about old Michaelmas
6	Snipe	*Gallinago minor*	Some snipes constantly breed with us
7	Jack snipe	*Gallinago minima*	—
8	Wood-pigeon	*Oenas*	Seldom appears till late: not in such plenty as formerly
9	Wild-swan	*Cygnus ferus*	On some large waters
10	Wild-goose	*Anser ferus*	—
11	Wild-duck	*Anas torquata minor*	On our lakes and streams
12	Pochard	*Anas fera fusca*	On our lakes and streams
13	Wigeon	*Penelope*	On our lakes and streams
14	Teal,	*Querquedula*	On our lakes and streams; breeds with us in Wolmer Forest
15	Cross-beak	*Coccothraustes*	These are only wanderers that appear occasionally and are not observant of any regular migration
16	Gross-bill	*Loxia*	
17	Silk-tail	*Garrulus bohemicus*	

These birds, as they stand numerically, belong to the following Linnaean genera:

1, 2, 3,	*Turdus.*	9, 10, 11,	
4,	*Corvus*	12, 13, 14	*Anas*
5, 6, 7,	*Scolopax*	15, 16	*Loxia*
8,	*Columba*	17	*Ampelis*

Birds that sing in the night are but few.

Nightingale	*Luscinia*	'In shadiest covert hid' – MILTON
Woodlark	*Alauda arborea*	Suspended in mid air
Less reed-sparrow	*Passer arundicaceus minor*	Among reeds and willows

I should now proceed to such birds as continue to sing after Midsummer, but, as they are rather numerous, they would exceed the bounds of this paper: besides, as this is now the season for remarking on that subject, I am willing to repeat my observations on some birds concerning the continuation of whose song I seem at present to have some doubt.

I am, &c.

Stock-dove

Snipe

Letter 2 also to the Honourable Daines Barrington

Selborne, November 2nd, 1769

Dear Sir – When I did myself the honour to write to you about the end of last June on the subject of natural history, I sent you a list of the summer birds of passage which I have observed in this neighbourhood; and also a list of the winter birds of passage: I mentioned besides those softbilled birds that stay with us the winter through in the south of England, and those that are remarkable for singing in the night.

According to my proposal, I shall now proceed to such birds (singing birds strictly so called) as continue in full song till after Midsummer; and shall range them somewhat in the order in which they first begin to open as the spring advances.

		RAII NOMINA	
1	Woodlark	*Alauda arborea*	In January, and continues to sing through all the summer and autumn
2	Song-thrush	*Turdus simpliciter dictus*	In February and on to August: re-assume their song in autumn
3	Wren	*Passer troglodytes*	All the year, hard frost excepted
4	Redbreast	*Rubecula*	Ditto
5	Hedge-sparrow	*Curruca*	Early in February to July 10
6	Yellowhammer	*Emberiza flava*	Early in February, and on through July to August 21
7	Skylark	*Alauda vulgaris*	In February, and on to October
8	Swallow	*Hirundo domestica*	From April to September
9	Black-cap	*Atricapilla*	Beginning of April to July 13
10	Titlark	*Alauda pratorum*	From middle of April to July 16
11	Blackbird	*Merula vulgaris*	Sometimes in February and March, and so on to July 23, re-assumes in autumn
12	Whitethroat	*Ficedulae affinis*	In April, and on to July 23
13	Goldfinch	*Carduelis*	April, and through to September 16
14	Greenfinch	*Chloris*	On to July and August 2
15	Less reed-sparrow	*Passer arundinaceus minor*	May on to beginning of July
16	Common linnet	*Linaria vulgaris*	Breeds and whistles on till August: re-assumes its note when they begin to congregate in October, and again early before the flocks separate

Birds that cease to be in full song, and are usually silent at or before Midsummer:

17	Middle willow wren	*Regulus non cristatus*	Middle of June: begins in April
18	Redstart	*Ruticilla*	Ditto: begins in May
19	Chaffinch	*Fringilla*	Beginning of June: sings first in February
20	Nightingale	*Luscinia*	Middle of June: sings first in April

Birds that sing for a short time, and very early in the spring:

21	Missel-bird,	*Turdus viscivorus*	January 2, 1770, in February. Is called in Hampshire and Sussex the storm-cock, because its song is supposed to forbode windy wet weather: it is the largest singing bird we have.
22	Great titmouse, or ox-eye	*Fringillago*	In February, March, April: reassumes for a short time in September

Birds that have somewhat of a note or song, and yet are hardly to be called singing birds:

23	Golden-crowned wren	*Regulus cristatus*	Its note as minute as its person: frequents the tops of high oaks and firs: the smallest British bird
24	Marsh-titmouse	*Parus palustris*	Haunts great woods: two harsh sharp notes
25	Small willow-wren	*Regulus non cristatus*	Sings in March and on to September
26	Largest ditto	*Ditto.*	*Cantat voce stridula locustae*; from end of April to August
27	Grass hopper lark	*Alauda minima voce locustae*	Chirps all night, from the middle of April to the end of July

28	Martin	*Hirundo agrestis*	All the breeding time: from May to September
29	Bullfinch	*Pyrrhula*	—
30	Bunting	*Emberiza alba.*	From the end of January to July

All singing birds, and those that have any pretensions to song, not only in Britain, but perhaps the world through, come under the Linnaean *ordo* of *Passeres*.

The above-mentioned birds, as they stand numerically, belong to the following Linnaean genera:

1, 7, 10, 27	*Alauda*	6, 30	*Emberiza*
2, 11, 21	*Turdus*	8, 28	*Hirundo*
3, 4, 5, 9, 12, 15, 17, 18, 20, 23, 25, 26	*Motacilla*	13, 16, 19	*Fringilla*
		22, 24	*Parus*
		14, 29	*Loxia*

Birds that sing as they fly are but few:

Skylark	*Alauda vulgaris*	Rising, suspended, and falling
Titlark	*Alauda pratorum*	In its descent: also sitting on trees, and walking on the ground
Woodlark	*Alauda arborea*	Suspended: in hot summer nights all night long
Blackbird	*Merula*	Sometimes from bush to bush
Whitethroat	*Ficeduiae affinis*	Uses when singing on the wing odd jerks and gesticulations
Swallow	*Hirundo domestica*	In soft sunny weather
Wren	*Passer troglodytes*	Sometimes from bush to bush

Birds that breed most early in these parts:

Raven	*Corvus*	Hatches in February and March
Song-thrush	*Turdus*	In March
Blackbird	*Merula*	In March
Rook	*Cornix frugilega*	Builds the beginning of March
Woodlark	*Alauda arborea*	Hatches in April
Ring-dove	*Palumbus torquatus*	Lays the beginning of April

All birds that continue in full song till after Midsummer appear to me to breed more than once.

Most kinds of birds seem to me to be wild and shy somewhat in proportion to their bulk; I mean in this island, where they are much pursued and annoyed; but in Ascension Island, and many other desolate places, mariners have found fowls so unacquainted with an human figure, that they would stand still to be taken; as is the case with boobies, &c. As an example of what is advanced, I remark that the gold-crested wren (the smallest British bird) will stand unconcerned till you come within three or four yards of it, while the bustard (*Otis*), the largest British land fowl, does not care to admit a person within so many furlongs.

I am, &c.

Yellowhammer

Letter 3 also to the Hon. Daines Barrington

Selborne, January 15th, 1770

Dear Sir – It was no small matter of satisfaction to me to find that you were not displeased with my little *methodus* of birds. If there was any merit in the sketch, it must be owing to its punctuality. For many months I carried a list in my pocket of the birds that were to be remarked, and, as I rode or walked about my business, I noted each day the continuance or omission of each bird's song; so that I am as sure of the certainty of my facts as a man can be of any transaction whatsoever.

I shall now proceed to answer the several queries which you put in your two obliging letters, in the best manner that I am able. Perhaps Eastwick, and its environs, where you heard so very few birds, is not a woodland country, and therefore not stocked with such songsters. If you will cast your eye on my last letter, you will find that many species continue to warble after the beginning of July.

The titlark and yellowhammer breed late, the latter very late; and therefore it is no wonder that they protract their song: for I lay it down as a maxim in ornithology, that as long as there is any incubation going on there is music. As to the redbreast and wren, it is well known to the most incurious observer that they whistle the year round, hard frost excepted; especially the latter.

It is not in my power to procure you a black-cap, or a less reedsparrow, or sedge-bird, alive. As the first is undoubtedly, and the last, as far as I can yet see, a summer bird of passage, they would require more nice and curious management in a cage than I should be able to give them: they are both distinguished songsters. The note of the former has such a wild sweetness that it always brings to my mind those lines in a song in *As You Like It*:

> And tune his merry note
> Unto the *wild* bird's throat.
>
> SHAKESPEARE

The latter has a surprising variety of notes resembling the song of several other birds; but then it has also an hurrying manner, not at all to its advantage: it is notwithstanding a delicate polyglot.

It is new to me that titlarks in cages sing in the night; perhaps only caged ones do so. I once knew a tame redbreast in a cage that always sang as long as candles were in the room; but in their wild state no one supposes they sing in the night.

I should be almost ready to doubt the fact, that there are to be seen much fewer birds in July than in any former month, notwithstanding so many young are hatched daily. Sure I am that it is far otherwise with respect to the swallow tribe, which increases prodigiously as the summer advances: and I saw at the time mentioned, many hundreds of young wagtails on the banks of the Cherwell, which almost covered the meadows. If the matter appears as you say in the other species, may it not be owing to the dams being engaged m incubation, while the young are concealed by the leaves?

Many times have I had the curiosity to open the stomachs of woodcocks and snipes; but nothing ever occurred that helped to explain to me what their subsistence might be: all that I could ever find was a soft mucus, among which lay many pellucid small gravels.

I am, &c.

Cuckoo

Letter 4 also to the Hon. Daines Barrington

Selborne, February 19th, 1770

Dear Sir – Your observation that 'the cuckoo does not deposit its egg indiscriminately in the nest of the first bird that comes in its way, but probably looks out a nurse in some degree congenerous, with whom to intrust its young', is perfectly new to me; and struck me so forcibly, that I naturally fell into a train of thought that led me to consider whether the fact was so, and what reason there was for it. When I came to recollect and inquire, I could not find that any cuckoo had ever been seen in these parts, except in the nest of the wagtail, the hedge-sparrow, the titlark, the whitethroat, and the redbreast, all soft-billed insectivorous birds. The excellent Mr Willughby mentions the nest of the *Palumbus* (ring-dove), and of the *fringilla* (chaffinch), birds that subsist on acorns and grains, and such hard food: but then he does not mention them as of his own knowledge; but says afterwards that he saw himself a wagtail feeding a cuckoo. It appears hardly possible that a soft-billed bird should subsist on the same food with the hard-billed: for the former have thin membranaceous stomachs suited to their soft food; while the latter, the granivorous tribe, have strong muscular gizzards which, like mills, grind, by the help of small gravels and pebbles, what is swallowed. This proceeding of the cuckoo, of dropping its eggs as it were by chance, is such a monstrous outrage on maternal affection, one of the first great dictates of nature; and such a violence on instinct; that, had it only been related of a bird in the Brazils, or Peru, it would never have merited our belief. But yet, should it farther appear that this simple bird, when divested of that natural στοργὴ that seems to raise the kind in general above themselves, and inspire them with extraordinary degrees of cunning and address, may be still endued with a more enlarged faculty of discerning what species are suitable and congenerous nursing-mothers for its disregarded eggs and young, and may deposit them only under their care, this would be adding wonder to wonder, and instancing, in a fresh manner, that the methods of Providence are not subjected to any mode or rule, but astonish us in new lights, and in various and changeable appearances.[1]

What was said by a very ancient and sublime writer concerning the defect of natural affection in the ostrich, may be well applied to the bird we are talking of:

> She is hardened against her young ones, as though they were not hers:
> Because God hath deprived her of wisdom, neither hath he imparted to her understanding.[2]

Query. Does each female cuckoo lay but one egg in a season, or does she drop several in different nests according as opportunity offers?[3]

I am, &c.

1 Job xxxix: 16, 17.

2 The cuckoo lays its eggs for the most part in the nests of birds much smaller than itself. This is probably in order that the young cuckoo may be markedly stronger than its fellow-nestlings, and so able to oust its unhappy little foster-brothers from the nest when necessary. It is quite true that cuckoos lay in the nests of chaffinches; but there is no such objection to this procedure as White supposes: for all the finches, as well as some other hard-billed birds. though they subsist in the adult stage on grains and acorns, feed their callow young upon grubs and caterpillars. It is not likely, on the other hand, that the cuckoo would lay in a ring-dove's nest, because the ring-dove would notice the marked difference in size, and the young ring-doves would also be quite as strong as the young cuckoo, better able to oust it than it would be to oust them. ED.

3 The cuckoo lays several eggs yearly, dropping them about in different nests as chances offer. ED.

Linnet

Letter 5 also to the Hon. Daines Barrington

Selborne, April 12th, 1770

Dear Sir – I heard many birds of several species sing last year after Midsummer; enough to prove that the summer solstice is not the period that puts a stop to the music of the woods. The yellow-hammer no doubt persists with more steadiness than any other; but the woodlark, the wren, the redbreast, the swallow, the whitethroat, the goldfinch, the common linnet, are all undoubted instances of the truth of what I advanced.

If this severe season does not interrupt the regularity of the summer migrations, the blackcap will be here in two or three days. I wish it was in my power to procure you one of those songsters; but I am no birdcatcher, and so little used to birds in a cage, that I fear if I had one it would soon die for want of skill in feeding.

Was your reed-sparrow, which you kept in a cage, the thick-billed reed-sparrow of the *Zoology*, p. 320; or was it the less reed-sparrow of Ray, the sedge-bird of Mr Pennant's last publication, p. 16?

As to the matter of long-billed birds growing fatter in moderate frosts, I have no doubt within myself what should be the reason. The thriving at those times appears to me to arise altogether from the gentle check which the cold throws upon insensible perspiration. The case is just the same with blackbirds, &c.; and farmers and warreners observe, the first, that their hogs fat more kindly at such times, and the latter that their rabbits are never in such good case as in a gentle frost. But when frosts ate severe, and of long continuance, the case is soon altered; for then a want of food soon overbalances the repletion occasioned by a checked perspiration. I have observed, moreover, that some human constitutions are more inclined to plumpness in winter than in summer.

When birds come to suffer by severe frost, I find that the first that fail and die are the redwing-fieldfares, and then the song-thrushes.

You wonder, with good reason, that the hedge-sparrows, &c., can be induced at all to sit on the egg of the cuckoo without being scandalized at the vast disproportionate size of the supposititious egg; but the brute creation, I suppose, have very little idea of size, colour, or number. For the common hen, I know, when the fury of incubation is on her, will sit on a single shapeless stone instead of a nest full of eggs that have been withdrawn: and, moreover, a hen-turkey, in the same circumstances, would sit on in the empty nest till she perished with hunger.[1]

I think the matter might easily be determined whether a cuckoo lays one or two eggs, or more, in a season, by opening a female during the laying time. If more than one was come down out of the ovary, and advanced to a good size, doubtless then she would that spring lay more than one.[2]

I will endeavour to get a hen, and to examine.

Your supposition that there may be some natural obstruction in singing birds while they are mute, and that when this is removed the song recommences, is new and bold; I wish you could discover some good grounds for this suspicion.

I was glad you were pleased with my specimen of the*caprimulgus*, or fern-owl; you were, I find, acquainted with the bird before.

1 As a matter of fact, the egg of the cuckoo is scarcely larger than that of the hedge-sparrows and chaffinches in whose nest the mother-bird lays: but the young cuckoo is very voracious, and therefore soon outgrows its small foster-brothers. Cuckoos have not been observed to lay in the nests of birds whose eggs are larger than their own. Their most common host, I think, is the meadow-pipit. ED.

2 This is physiologically incorrect. Only one ovum is contained in the oviduct of any bird at one time; but the cuckoo, as already noted, does lay several eggs in each season. ED.

When we meet I shall be glad to have some conversation with you concerning the proposal you make of my drawing up an account of the animals in this neighbourhood. Your partiality towards my small abilities persuades you, I fear, that I am able to do more than is in my power: for it is no small undertaking for a man unsupported and alone to begin a natural history from his own autopsia! Though there is endless room for observation in the field of nature, which is boundless, yet investigation (where a man endeavours to be sure of his facts) can make but slow progress; and all that one could collect in many years would go into a very narrow compass.

Some extracts from your ingenious 'Investigations of the Difference between the Present Temperature of the Air in Italy', &c., have fallen in my way; and gave me great satisfaction: they have removed the objections that always arose in my mind whenever I came to the passages which you quote. Surely the judicious Virgil, when writing a didactic poem for the region of Italy, could never think of describing freezing rivers, unless such severity of weather pretty frequently occurred.

PS – Swallows appear amidst snows and frost.

Alton from Windmill Hill

Reed sparrow

Letter 6 also to the Hon. Daines Barrington

Selborne, May 21st, 1770

Dear Sir – The severity and turbulence of last month so interrupted the regular process of summer migration, that some of the birds do but just begin to show themselves, and others are apparently thinner than usual; as the whitethroat, the black-cap, the red-start, the flycatcher. I well remember that after the very severe spring in the year 1739–40, summer birds of passage were very scarce. They come probably hither with a south-east wind, or when it blows between those points; but in that unfavourable year the winds blowed the whole spring and summer through from the opposite quarters. And yet amidst all these disadvantages two swallows, as I mentioned in my last, appeared this year as early as the eleventh of April amidst frost and snow; but they withdrew again for a time.

I am not pleased to find that some people seem so little satisfied with Scopoli's new publication;[1] there is room to expect great things from the hands of that man, who is a good naturalist: and one would think that an history of the birds of so distant and southern a region as Carniola would be new and interesting. I could wish to see

1 This work he calls his *Annus Primus Historico-Naturalis*.

that work, and hope to get it sent down. Dr Scopoli is physician to the wretches that work in the quicksilver mines of that district.

When you talked of keeping a reed-sparrow, and giving it seeds, I could not help wondering; because the reed-sparrow which I mentioned to you (*Passer arundinaceus minor Raii*) is a soft-billed bird; and most probably migrates hence before winter; whereas the bird you kept (*Passer torquatus Raii*) abides all the year, and is a thick-billed bird.[2] I question whether the latter be much of a songster; but in this matter I want to be better informed. The former has a variety of hurrying notes, and sings all night. Some part of the song of the former, I suspect, is attributed to the latter. We have plenty of the soft-billed sort; which Mr Pennant had entirely left out of his *British Zoology*, till I reminded him of his omission. See *British Zoology* last published, p. 16.[3]

I have somewhat to advance on the different manners in which different birds fly and walk; but as this is a subject that I have not enough considered, and is of such a nature as not to be contained in a small space, I shall say nothing further about it at present.[4]

No doubt the reason why the sex of birds in their first plumage is so difficult to be distinguished is, as you say, 'because they are not to pair and discharge their parental functions till the ensuing spring'. As colours seem to be the chief external sexual distinction in many birds, these colours do not take place till sexual attachments begin to obtain. And the case is the same in quadrupeds; among whom, in their younger days, the sexes differ but little; but, as they advance to maturity, horns and shaggy manes, beards and brawny necks, &c. &c., strongly discriminate the male from the female. We may instance still farther in our own species, where a beard and stronger features are usually characteristic of the male sex: but this sexual diversity does not take place in earlier life; for a beautiful youth shall be so like a beautiful girl that the difference shall not be discernible:

> *Quem si puellarum insereres choro,*
> *Mire sagaces falleret hospites*
> *Discrimen obscurum, solutis*
> *Crinibus, ambiguoque cultu.*
>
> HORACE

2 See Letter 25 to Mr Pennant.
3 See Letter 42 to Mr Barrington.
4 The bird here alluded to is the reed-bunting. ED.

Letter 7 also to the Hon. Daines Barrington

Ringmer, near Lewes, October 8th, 1770

Dear Sir, I am glad to hear that Kuckalm is to furnish you with the birds of Jamaica; a sight of the hirundines of that hot and distant island would be a great entertainment to me.

The *Anni* of Scopoli are now in my possession; and I have read *Annus Primum* with satisfaction; for though some parts of this work are exceptionable, and he may advance some mistaken observations, yet the ornithology of so distant a country as Carniola is very curious. Men that undertake only one district are much more likely to advance natural knowledge that those that grasp at more than they can possibly be acquainted with: every kingdom, every province, should have its own monographer.

The reason perhaps why he mentions nothing of Ray's *Ornithology* may be the extreme poverty and distance of his country, into which the works of our great naturalist may have never yet found their way. You have doubts, I know, whether this *Ornithology* is genuine, and really the work of Scopoli; as to myself, I think I discover strong tokens of authenticity; the style corresponds with that of his

Entomology; and his characters of his Ordines and Genera are many of them new, expressive, and masterly. He has ventured to alter some of the Linnaean genera with sufficient show of reason.

It might perhaps be mere accident that you saw so many swifts and no swallows at Staines; because, in my long observations of those birds, I never could discover the least degree of rivalry or hostility between the species.[1]

Ray remarks that birds of the *gallinae* order, as cocks and hens, partridges, and pheasants, &c., are *pulveratrices*, such as dust themselves, using that method of cleansing their feathers, and ridding themselves of their vermin. As far as I can observe, many birds that dust themselves never wash; and I once thought that those birds that wash themselves would never dust; but here I find myself mistaken; for common house-sparrows are great pulveratrices, being frequently seen grovelling and wallowing in dusty roads; and yet they are great washers. Does not the skylark dust?[2]

Query. Might not Mahomet and his followers take one method of purification from these pulveratrices? because I find from travellers of credit, that if a strict Mussulman is journeying in a sandy desert where no water is to be found, at stated hours he strips off his clothes, and most scrupulously rubs his body over with sand or dust.

A countryman told me he had found a young fern-owl in the nest of a small bird on the ground; and that it was fed by the little bird. I went to see this extraordinary phenomenon, and found that it was a young cuckoo hatched in the nest of a titlark; it was become vastly too big for its nest, appearing

> . . . *in tenue re*
> *Majores pennas nido extendisse*

and was very fierce and pugnacious, pursuing my finger, as I teased it, for many feet from the nest, and sparring and buffeting with its wings like a game-cock. The dupe of a dam appeared at a distance, hovering about with meat in its mouth, and expressing the greatest solicitude.

In July I saw several cuckoos skimming over a large pond; and

1 White, I think, is here in error. Both birds feed on the same sort of insects, which they catch on the wing, and several observers have noticed the swift flying, or swoophng in a hostile manner at swallows and house-martins. ED.

2 White is correct in this; skylarks wash themselves by dusting. ED.

found, after some observation, that they were feeding on the *Libellulae*, or dragonflies; some of which they caught as they settled on the weeds, and some as they were on the wing. Notwithstanding what Linnaeus says, I cannot be induced to believe that they are birds of prey.

This district affords some birds, that are hardly ever heard of at Selborne. In the first place considerable flocks of cross-beaks (*Loxiae curvirostrae*) have appeared this summer in the pine-groves belonging to this house; the water-ousel is said to haunt the mouth of the Lewes river, near Newhaven; and the Cornish chough builds, I know, all along the chalky cliffs of the Sussex shore.

I was greatly pleased to see little parties of ring-ousels (my newly discovered migrators) scattered, at intervals, all along the Sussex downs, from Chichester to Lewes. Let them come from whence they will, it looks very suspicious that they are cantoned along the coast in order to pass the channel when severe weather advances. They visit us again in April, as it should seem, in their return; and are not to be found in the dead of winter. It is remarkable that they are very tame, and seem to have no manner of apprehensions of danger from a person with a gun. There are bustards on the wide downs near Brighthelmstone [Brighton]. No doubt you are acquainted with the Sussex downs; the prospects and rides round Lewes are most lovely.

As I rode along near the coast I kept a very sharp look-out in the lanes and woods, hoping I might, at this time of the year, have discovered some of the summer short-winged birds of passage crowding towards the coast in order for their departure: but it was very extraordinary that I never saw a redstart, white-throat, black-cap, uncrested wren, flycatcher, &c. And I remember to have made the same remark in former years, as I usually come to this place annually about this time. The birds most common along the coast, at present, are the stone-chatters, winchats, buntings, linnets, some few wheat-ears, titlarks, &c. Swallows and housemartins abound yet, induced to prolong their stay by this soft, still, dry season.

A land tortoise, which has been kept for thirty years in a little walled court belonging to the house where I now am visiting, retires under ground about the middle of November, and comes forth again about the middle of April. When it first appears in the spring it discovers very little inclination towards food; but in the height of summer grows voracious; and then as the summer declines its appetite declines; so that for the last six weeks in autumn it hardly eats at all. Milky plants, such as lettuces, dandelions, sowthistles, are

The Butcher's shop

its favourite dish. In a neighbouring village one was kept till by tradition it was supposed to be an hundred years old. An instance of vast longevity in such a poor reptile![3]

3 Still older instances are now on record. ED.

Crossbill

Letter 8 also to the Hon. Daines Barrington

Selborne, December 20th, 1770

Dear Sir – The birds that I took for aberdavines were reed-sparrows (*Passeres torquati*).

There are doubtless many home internal migrations within this kingdom that want to be better understood: witness those vast flocks of hen chaffinches that appear with us in the winter without hardly any cocks among them. Now was there a due proportion of each sex, it should seem very improbable that any one district should produce such numbers of these little birds; and much more when only one-half of the species appears; therefore we may conclude that the *Fringillae coelebes*, for some good purposes, have a peculiar migration of their own in which the sexes part. Nor should it seem so wonderful that the intercourse of sexes in this species of bird should be interrupted in winter; since in many animals, and particularly in bucks and does, the sexes herd separately, except at the season when commerce is necessary for the continuance of the breed. For this matter of the chaffinches see *Fauna Suecica*, p. 85, and *Systema Naturae*, p. 318. I see every winter vast flights of hen chaffinches, but none of cocks.

Your method of accounting for the periodical motions of the British singing-birds, or birds of flight, is a very probable one; since the matter of food is a great regulator of the actions and proceedings of the brute creation; there is but one that can be set in competition with it, and that is love. But I cannot quite acquiesce with you in one circumstance when you advance that 'when they have thus feasted, they again separate into small parties of five or six, and get the best fare they can within a certain district, having no inducement to go in quest of fresh-turned earth'. Now if you mean that the business of congregating is quite at an end from the conclusion of wheat sowing to the season of barley and oats, it is not the case with us; for larks and chaffinches, and particularly linnets, flock and congregate as much in the very dead of winter as when the husbandman is busy with his ploughs and harrows.

Sure there can be no doubt but that woodcocks and fieldfares leave us in the spring, in order to cross the seas, and to retire to some districts more suitable to the purpose of breeding. That the former pair before they retire, and that the hens are forward with egg, I myself, when I was a sportsman, have often experienced. It cannot indeed be denied but that now and then we hear of a woodcock's nest, or young birds, discovered in some part or other of this island; but then they are always mentioned as rarities, and somewhat out of the common course of things; but as to redwings and fieldfares, no sportsman or naturalist has ever yet, that I could hear, pretended to have found the nest or young of those species in any part of these kingdoms. And I the more admire at this instance as extraordinary, since, to all appearance, the same food in summer as well as in winter might support them here which maintains their congeners, the blackbirds and thrushes, did they choose to stay the summer through. From hence it appears that it is not food alone which determines some species of birds with regard to their stay or departure. Fieldfares and redwings disappear sooner or later according as the warm weather comes on earlier or later. For I well remember, after that dreadful winter 1739–40, that cold north-east winds continued to blow on through April and May, and that these kind of birds (what few remained of them) did not depart as usual, but were seen lingering about till the beginning of June.

The best authority that we can have for the nidification of the birds above-mentioned in any district, is the testimony of faunists that have written professedly the natural history of particular countries. Now as to the fieldfare, Linnaeus, in his *Fauna Suecica*, says of it, that '*maximis in arboribus nidificat*'; and of the redwing he says, in the same place, that '*nidificat in mediis arbusculis, sive sepibus: ova sex caeruleo-viridia maculis nigris variis*'. Hence we may be assured that fieldfares and redwings breed in Sweden. Scopoli says, in his *Annus Primus*, of the woodcock, that '*nupta ad nos venit circa aequinoctium vernale*'; meaning in Tyrol, of which he is a native. And afterwards he adds, '*nidificat in paludibus alpinis: ova ponit* 3–5'. It does not appear from Kramer that woodcocks breed at all in Austria; but he says, '*Avis haec septentrionalium provinciarum aestivo tempore incola est; ubi plerumque nidificat. Appropinquante hyeme australiores provincias petit; hinc circa plenilunium mensis Octobris plerumque Austriam transmigrat. Tunc rursus circa plenilunium potissimum mensis Martii per Austriam matrimonio juncta ad septentrionales provincias redit.*' For the whole passage (which I have abridged) see *Elenchus*, &c. p. 351 . This seems to be a full proof of the migration of woodcocks; though little is proved concerning the place of breeding.

PS – There fell in the county of Rutland, in three weeks of this present very wet weather, seven inches and a half of rain, which is more than has fallen in any three weeks for these thirty years past in that part of the world. A mean quantity in that county for one year is twenty inches and a half.

Gracious Street

Letter 9 also to the Hon. Daines Barrington

Fyfield near Andover, February 12th, 1772

Dear Sir – You are, I know, no great friend to migration; and the well-attested accounts from various parts of the kingdom seem to justify you in your suspicions, that at least many of the swallow kind do not leave us in the winter, but lay themselves up like insects and bats, in a torpid state, and slumber away the more uncomfortable months till the return of the sun and fine weather awakens them!

But then we must not, I think, deny migration in general; because migration certainly does subsist in some places, as my brother in Andalusia has fully informed me. Of the motions of these birds he has ocular demonstration, for many weeks together, both spring and

1 This letter is in answer to an essay of Barrington's, published in his *Miscellanies*, p. 174, 'On the Periodical Appearing and Disappearing of Certain Birds at Different Times of the Year'. In that paper Barrington argues against the probability of periodical migration; and White here meets many of his rather fanciful difficulties and objections. ED.

fall; during which periods myriads of the swallow kind traverse the Straits from north to south, and from south to north, according to the season. And these vast migrations consist not only of hirundines but of bee-birds, hoopoes, *Oro pendolos*, or golden thrushes, &c. &c., and also of many of our soft-billed summer birds of passage; and moreover of birds which never leave us, such as all the various sorts of hawks and kites. Old Belon, two hundred years ago, gives a curious account of the incredible armies of hawks and kites which he saw in the springtime traversing the Thracian Bosphorus from Asia to Europe. Besides the above-mentioned, he remarks that the procession is swelled by whole troops of eagles and vultures.

Now it is no wonder that birds residing in Africa should retreat before the sun as it advances, and retire to milder regions, and especially birds of prey, whose blood being heated with hot animal food, are more impatient of a sultry climate; but then I cannot help wondering why kites and hawks, and such hardy birds as are known to defy all the severity of England, and even of Sweden and all north Europe, should want to migrate from the south of Europe, and be dissatisfied with the winters of Andalusia.[2]

It does not appear to me that much stress may be laid on the difficulty and hazard that birds must run in their migrations, by reason of vast oceans, cross winds, &c.; because, if we reflect, a bird may travel from England to the Equator without launching out and exposing itself to boundless seas, and that by crossing the water at Dover, and again at Gibraltar. And I with the more confidence advance this obvious remark, because my brother has always found that some of his birds, and particularly the swallow kind, are very sparing of their pains in crossing the Mediterranean; for when arrived at Gibraltar they do not:

> Rang'd in figure wedge their way,
> . . . And set forth
> Their airy caravan high over seas
> Flying, and over lands with mutual wing
> Eating their flight:
>
> MILTON

but scout and hurry along in little detached parties of six or seven in a company; and sweeping low, just over the surface of the land and water, direct their course to the opposite continent at the narrowest

2 It is curious to see how often and how persistently our great naturalist returns to this question of migration. ED.

passage they can find. They usually slope across the bay to the south-west, and so pass over opposite to Tangier, which, it seems, is the narrowest space.

In former letters we have considered whether it was probable that woodcocks in moonshiny nights cross the German ocean from Scandinavia. As a proof that birds of less speed may pass that sea, considerable as it is, I shall relate the following incident, which, though mentioned to have happened so many years ago, was strictly matter of fact: – As some people were shooting in the parish of Trotton, in the county of Sussex, they killed a duck in that dreadful winter, 1708–9, with a silver collar about its neck,[3] on which were engraven the arms of the king of Denmark. This anecdote the rector of Trotton at that time has often told to a near relation of mine; and, to the best of my remembrance, the collar was in the possession of the rector.

At present I do not know anybody near the seaside that will take the trouble to remark at what time of the moon woodcocks first come; if I lived near the sea myself I would soon tell you more of the matter. One thing I used to observe when I was a sportsman, that there were times in which woodcocks were so sluggish and sleepy that they would drop again when flushed just before the spaniels, nay, just at the muzzle of a gun that had been fired at them; whether this strange laziness was the effect of a recent fatiguing journey I shall not presume to say.

Nightingales not only never reach Northumberland and Scotland, but also, as I have been always told, Devonshire and Cornwall. In those last two counties we cannot attribute the failure of them to the want of warmth; the defect in the west is rather a presumptive argument that these birds come over to us from the continent at the narrowest passage, and do not stroll so far westward.

Let me hear from your own observation whether skylarks do not dust. I think they do; and if they do, whether they wash also.

The *Alauda pratensis* of Ray was the poor dupe that was educating the booby of a cuckoo mentioned in my letter of October last.

Your letter came too late for me to procure a ring-ousel for Mr Tunstal during their autumnal visit; but I will endeavour to get him one when they call on us again in April. I am glad that you and that gentleman saw my Andalusian birds; I hope they answered your expectation. Royston, or grey crows, are winter birds that come

3 I have read a like anecdote of a swan.

much about the same time with the woodcock; they, like the fieldfare and redwing, have no apparent reason for migration; for as they fare in the winter like their congeners, so might they in all appearance in the summer. Was not Tenant, when a boy, mistaken? did he not find a missel-thrush's nest, and take it for the nest of a fieldfare?

The stock-dove, or wood-pigeon, *Oenas Raii*, is the last winter bird of passage which appears with us; it is not seen till towards the end of November: about twenty years ago they abounded in the district of Selborne; and strings of them were seen morning and evening that reached a mile or more; but since the beechen woods have been greatly thinned they are much decreased in number. The ring-dove, *Palumbus Raii*, stays with us the whole year, and breeds several times through the summer.

Before I received your letter of October last I had just remarked in my journal that the trees were unusually green. This uncommon verdure lasted on late into November; and may be accounted for from a late spring, a cool and moist summer; but more particularly from vast armies of chafers, or tree-beetles, which, in many places, reduced whole woods to a leafless naked state. These trees shot again at Midsummer, and then retained their foliage till very late in the year.

My musical friend, at whose house I am now visiting, has tried all the owls that are his near neighbours with a pitch-pipe set at concert pitch, and finds they all hoot in B flat. He will examine the nightingales next spring.

I am, &c. &c.

The Yew Tree

Letter 10 also to the Hon. Daines Barrington

Selborne, August 1st, 1771

Dear Sir – From what follows, it will appear that neither owls nor cuckoos keep to one note. A friend remarks that many (most)*of* his owls hoot in B flat; but that one went almost half a note below A. The pipe he tried their notes by was a common half-crown pitch-pipe, such as masters use for tuning of harpsichords; it was the common London pitch.

A neighbour of mine, who is said to have a nice ear, remarks that the owls about this village hoot in three different keys, in G flat, or F sharp, in B flat and A flat. He heard two hooting to each other, the one in A flat and the other in B flat.

Query: Do these different notes proceed from different species, or only from various individuals? The same person finds upon trial that the note of the cuckoo (of which we have but one species) varies in different individuals; for, about Selborne wood, he found they were mostly in D: he heard two sing together, the one in D, the other in D sharp, who made a disagreeable concert: he afterwards heard one in D sharp, and about Wolmer forest some in C. As to nightingales, he says that their notes are so short, and their transitions so rapid, that he cannot well ascertain their key. Perhaps in a cage, and in a room, their notes may be more distinguishable. This person has tried to settle the notes of a swift, and of several other small birds, but cannot bring them to any criterion.

As I have often remarked that redwings are some of the first birds that suffer with us in severe weather, it is no wonder at all that they retreat from Scandinavian winters: and much more the *ordo*, of *grallae*, who, all to a bird, forsake the northern parts of Europe at the approach of winter. '*Grallae tanquam conjurate unanimiter in fugam se conjiciunt; ne earum unicam quidem inter nos habitantem invenire possimus; ut enim aestate in australibus degere nequeunt ob defectum lumbricorum, terramque siccam; ita nec in frigidis ob eandem causam,*' says Ekmarck the Swede, in his ingenious little treatise called 'Migrationes Avium', which by all means you ought to read while your thoughts run on the subject of migration. See *Amoenitates Academicae*, vol. iv., p. 565.

Birds may be so circumstanced as to be obliged to migrate in one country, and not in another: but the *grallae* (which procure their food from marshy and boggy grounds), must in winter forsake the more northerly parts of Europe, or perish for want of food![1]

I am glad you are making inquiries from Linnaeus concerning the woodcock: it is expected of him that he should be able to account for the motions and manner of life of the animals of his own 'Fauna'.

Faunists, as you observe, are too apt to acquiesce in bare descriptions, and a few synonyms: the reason is plain; because all that may be done at home in a man's study, but the investigation of the life and conversation of animals is a concern of much more trouble and difficulty, and is not to be attained but by the active and inquisitive, and by those that reside much in the country.[2]

Foreign systematics[3] are, I observe, much too vague in their specific differences; which are almost universally constituted by one or two particular marks, the rest of the description running in general terms. But our countryman, the excellent Mr Ray, is the only describer that conveys some precise idea in every term or

1 The question of food-supply has far more to do with the migrations of birds than mere climate. ED.

2 In this passage White strikes the keynote of the modern school of natural history, the school which culminated in Darwin, and which is interested rather in the facts and problems of life than in mere classification. ED.

3 He means Systematists. ED.

word, maintaining his superiority over his followers and imitators in spite of the advantage of fresh discoveries and modern information.

At this distance of years it is not in my power to recollect at what period woodcocks used to be sluggish or alert when I was a sportsman: but, upon my mentioning this circumstance to a friend, he thinks he has observed them to be remarkably listless against snowy foul weather; if this should be the case, then the inaptitude for flying arises only from an eagerness for food; as sheep are observed to be very intent on grazing against stormy wet evenings.

I am, &c. &c.

Letter 11 also to the Hon. Daines Barrington

Selborne, February 8th, 1772

Dear Sir – When I ride about in the winter, and see such prodigious flocks of various kinds of birds, I cannot help admiring at these congregations, and wishing that it was in my power to account for those appearances almost peculiar to the season. The two great motives which regulate the proceedings of the brute creation are love and hunger; the former incites animals to perpetuate their kind; the latter induces them to preserve individuals; whether either of these should seem to be the ruling passion in the matter of congregating is to be considered. As to love, that is out of the question at a time of the year when that soft passion is not indulged: besides, during the amorous season, such a jealousy prevails between the male birds that they can hardly bear to be together in the same hedge or field. Most of the singing and elation of spirits of that time seem to me to be the effect of rivalry and emulation: and it is to this spirit of jealousy that I chiefly attribute the equal dispersion of birds in the spring over the face of the country.

Now as to the business of food: as these animals are actuated by instinct to hunt for necessary food, they should not, one would suppose, crowd together in pursuit of sustenance at a time when it is most likely to fail; yet such associations do take place in hard weather chiefly, and thicken as the severity increases. As some kind of self-interest and self-defence is no doubt the motive for the proceeding, may it not arise from the helplessness of their state in such rigorous seasons; as men crowd together, when under great calamities, though they know not why? Perhaps approximation may dispel some degree of cold; and a crowd may make each individual appear safer from the ravages of birds of prey and other dangers.

If I admire when I see how much congenerous birds love to congregate, I am the more struck when I see incongruous ones in such strict amity. If we do not much wonder to see a flock of rooks usually attended by a train of daws, yet it is strange that the former should so frequently have a flight of starlings for their satellites. Is it because rooks have a more discerning scent than their attendants, and can lead them to spots more productive of food? Anatomists say that rooks, by reason of two large nerves which run down between the eyes into the upper mandible, have a more delicate feeling in their beaks than other round-billed birds, and can grope for their meat when out of sight. Perhaps, then, their associates attend them on the motive of interest, as greyhounds wait on the motions of their finders; and as lions are said to do on the yelpings of jackals. Lapwings and starlings sometimes associate.

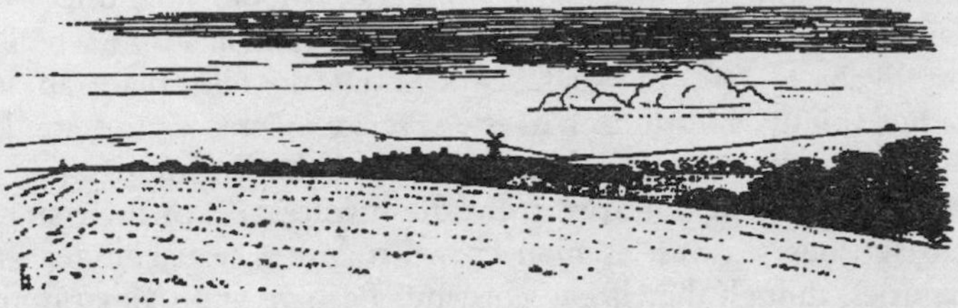

Lewes

Letter 12 also to the Hon. Daines Barrington

March 9th, 1772

Dear Sir – As a gentleman and myself were walking on the fourth of last November round the sea-banks at Newhaven, near the mouth of the Lewes river, in pursuit of natural knowledge, we were surprised to see three house-swallows gliding very swiftly by us. That morning was rather chilly, with the wind at north-west; but the tenor of the weather for some time before had been delicate, and the noons remarkably warm. From this incident, and from repeated accounts which I met with, I am more and more induced to believe that many of the swallow kind do not depart from this island, but lay themselves up in holes and caverns; and do, insect-like and batlike, come forth at mild times, and then retire again to their *latebrae*. Nor make I the least doubt but that, if I lived at Newhaven, Seaford, Brighthelmstone [Brighton], or any of those towns near the chalk cliffs of the Sussex coast, by proper observations I should see swallows stirring at periods of the winter when the noons were soft and inviting, and the sun warm and invigorating. And I am the more of this opinion from what I have remarked during some of our late springs, that though some swallows did make their appearance about the usual time, viz., the thirteenth or fourteenth of April, yet meeting with an harsh reception, and blustering cold north-east winds, they immediately withdrew, absconding for several days, till the weather gave them better encouragement.[1]

1 In all probability these swallows were either killed by the cold, or else returned southward to wait for warmer weather. I think the former. ED.

Letter 13 also to the Hon. Daines Barrington

April 12th, 1772

Dear Sir – While I was in Sussex last autumn my residence was at the village near Lewes, from whence I had formerly the pleasure of writing to you. On the first of November I remarked that the old tortoise, formerly mentioned, began first to dig the ground in order to the forming its hibernaculum, which it had fixed on just beside a great tuft of hepaticas. It scrapes out the ground with its forefeet, and throws it up over its back with its hind; but the motion of its legs is ridiculously slow, little exceeding the hour-hand of a clock: and suitable to the composure of an animal said to be a whole month in performing one feat of copulation. Nothing can be more assiduous than this creature night and day in scooping the earth, and forcing its great body into the cavity; but, as the noons of that season proved unusually warm and sunny, it was continually interrupted, and called forth by the heat in the middle of the day; and though I continued there till the thirteenth of November, yet the work remained unfinished. Harsher weather, and frosty mornings, would have quickened its operations. No part of its behaviour ever struck me more than the extreme timidity it always expresses with regard to rain; for though it has a shell that would secure it against the wheel of a loaded cart, yet does it discover as much solicitude about rain as a lady dressed in all her best attire, shuffling away on the first sprinklings, and running its head up in a corner. If attended to, it becomes an excellent weather-glass; for as sure as it walks elate, and as it were on tiptoe, feeding with great earnestness in a morning, so sure will it rain before night. It is totally a diurnal animal, and never pretends to stir after it becomes dark. The tortoise, like other reptiles, has an arbitrary stomach as well as lungs; and can refrain from eating as well as breathing for a great part of the year. When first awakened it eats nothing; nor again in the autumn before it retires: through the height of the summer it feeds voraciously, devouring all the food that comes in its way. I was much taken with its sagacity in discerning those that do it kind offices: for, as soon as the good old lady comes in sight who has waited on it for more than thirty years, it hobbles towards its benefactress with awkward alacrity; but remains inattentive to strangers. Thus not only 'the ox knoweth his owner, and the ass his master's crib',[1] but the most abject reptile and torpid of beings distinguishes the hand that feeds it, and is touched with the feelings of gratitude!

I am, &c. &c.

PS – In about three days after I left Sussex the tortoise retired into the ground under the hepatica.

1 Isaiah 1: 3.

Letter 14 also to the Hon. Daines Barrington

Selborne, March 26th, 1773

Dear Sir – The more I reflect on the στοργὴ of animals, the more I am astonished at its effects. Nor is the violence of this affection more wonderful than the shortness of its duration. Thus every hen is in her turn the virago of the yard, in proportion to the helplessness of her brood; and will fly in the face of a dog or a sow in defence of those chickens, which in a few weeks she will drive before her with relentless cruelty.

This affection sublimes the passions, quickens the invention, and sharpens the sagacity of the brute creation. Thus an hen, just become a mother, is no longer that placid bird she used to be, but with feathers standing on end, wings hovering, and clucking note, she runs about like one possessed. Dams will throw themselves in the way of the greatest danger in order to avert it from their progeny. Thus a partridge will tumble along before a sportsman in order to draw away the dogs from her helpless covey. In the time of nidification the most feeble birds will assault the most rapacious. All the hirundines of a village are up in arms at the sight of an hawk, whom they will persecute till he leaves that district. A very exact observer has often remarked that a pair of ravens nesting in the rock of Gibraltar would suffer no vulture or eagle to rest near the station, but would drive them from the hill with an amazing fury; even the blue thrush at the season of breeding would dart out from the clefts of the rocks to chase away the kestrel, or the sparrow-hawk. If you stand near the nest of a bird that has young, she will not be induced to betray them by an inadvertent fondness, but will wait about at a distance with meat in her mouth for an hour together.

Should I farther corroborate what I have advanced above by some anecdotes which I probably may have mentioned before in conversation, yet you will, I trust, pardon the repetition for the sake of the illustration.

The flycatcher of the *Zoology* (the *Stoparola* of Ray), builds every year in the vines that grow on the walls of my house. A pair of these little birds had one year inadvertently placed their nest on a naked bough, perhaps in a shady time, not being aware of the inconvenience that followed. But an hot sunny season coming on before the brood was half-fledged, the reflection of the wall became insupportable, and must inevitably have destroyed the tender young, had not affection suggested an expedient, and prompted the parent-birds to hover over the nest all the hotter hours, while with wings expanded, and mouths gaping for breath, they screened off the heat from their suffering offspring.

A farther instance I once saw of notable sagacity in a willow-wren, which had built in a bank in my fields. This bird a friend and myself had observed as she sat in her nest; but were particularly careful not to disturb her, though we saw she eyed us with some degree of jealousy. Some days after as we passed that way we were desirous of remarking how this brood went on; but no nest could be found, till I happened to take up a large bundle of long green moss, as it were, carelessly thrown over the nest in order to dodge the eye of an impertinent intruder.

A still more remarkable mixture of sagacity and instinct occurred to me one day as my people were pulling off the lining of an hotbed, in order to add some fresh dung. From out of the side of this bed leaped an animal with great agility that made a most grotesque figure; nor was it without great difficulty that it could be taken; when it proved to be a large white-bellied field-mouse with three or four young clinging to her teats by their mouths and feet. It was amazing that the desultory and rapid motions of this dam should not oblige her litter to quit their hold, especially when it appeared that they were so young as to be both naked and blind!

Selborne Street

To these instances of tender attachment, many more of which might be daily discovered by those that are studious of nature, may be opposed that rage of affection, that monstrous perversion of the στοργὴ, which induces some females of the brute creation to devour their young because their owners have handled them too freely, or removed them from place to place! Swine, and sometimes the more gentle race of dogs and cats, are guilty of this horrid and preposterous murder. When I hear now and then of an abandoned mother that destroys her offspring, I am not so much amazed; since reason perverted, and the bad passions let loose, are capable of any enormity; but why the parental feelings of brutes, that usually flow in one most uniform tenor, should sometimes be so extravagantly diverted, I leave to abler philosophers than myself to determine![1]

I am, &c.

1 It is now recognised that both in human beings and in the lower animals such a perversion of instinct is brought about by any violent and overwhelming emotion, fear, distress, or pain, affecting the mother immediately after the moment of maternity. It is an emotional period, and the emotions it rouses are easily turned into strange channels. ED.

White Owls

Letter 15 also to the Hon. Daines Barrington

Selborne, July 8th, 1773

Dear Sir – Some young men went down lately to a pond on the verge of Wolmer Forest to hunt flappers, or young wild-ducks, many of which they caught, and, among the rest, some very minute yet well-fledged wild-fowls alive, which upon examination I found to be teals. I did not know till then that teals ever bred in the south of England, and was much pleased with the discovery: this I look upon as a great stroke in natural history

We have had, ever since I can remember, a pair of white owls that constantly breed under the eaves of this church. As I have paid good attention to the manner of life of these birds during their season of breeding, which lasts the summer through, the following remarks may not perhaps be unacceptable – About an hour before sunset (for then the mice begin to run) they sally forth in quest of prey, and hunt all round the hedges of meadows and small enclosures for them, which seem to be their only food. In this irregular country we can stand on an eminence and see them beat the fields over like a setting-dog, and often drop down in the grass or corn. I have minuted these birds with my watch for an hour together, and have found that they return to their nest, the one or the other of them, about once in five minutes; reflecting at the same time on the adroitness that every animal is possessed of as far as regards the well-being of itself and offspring. But a piece of address, which they show when they return loaded, should not, I think, be passed over in silence. As they take their prey with their claws, so they carry it in their claws to their nest; but, as the feet are necessary in their ascent under the tiles, they constantly perch first on the roof of the chancel, and shift the mouse from their claws to their bill, that their feet may be at liberty to take hold of the plate on the wall as they are rising under the eaves.

White owls seem not (but in this I am not positive) to hoot at all; all that clamorous hooting appears to me to come from the wood kinds. The white owl does indeed snore and hiss in a tremendous manner; and these menaces well answer the intention of intimidating; for I have known a whole village up in arms on such an occasion, imagining the churchyard to be full of goblins and spectres. White owls also often scream horribly as they fly along; from this screaming probably arose the common people's imaginary species of screech-owl, which they superstitiously think attends the windows of dying persons. The plumage of the remiges of the wings of every species of owl that I have yet examined is remarkably soft and pliant. Perhaps it may be necessary that the wings of these birds should not make much resistance or rushing, that they may be enabled to steal through the air unheard upon a nimble and watchful quarry.

While I am talking of owls, it may not be improper to mention what I was told by a gentleman of the county of Wilts. As they were grubbing a vast hollow pollard-ash that had been the mansion of owls for centuries, he discovered at the bottom a mass of matter that at first he could not account for. After some examination he found that it was a congeries of the bones of mice (and perhaps of birds and bats) that had been heaping together for ages, being cast up in pellets out of the crops of many generations of inhabitants. For owls cast up the bones, fur, and feathers, of what they devour, after the manner of hawks. He believes, he told me, that there were bushels of this kind of substance.

When brown owls hoot, their throats swell as big as an hen's egg. I have known an owl of this species live a full year without any water. Perhaps the case may be the same with all birds of prey.

When owls fly they stretch out their legs behind them as a balance to their large heavy heads, for as most nocturnal birds have large eyes and ears they must have large heads to contain them. Large eyes I presume are necessary to collect every ray of light, and large concave ears to command the smallest degree of sound or noise.

I am, &c.

[It will be proper to premise here that the sixteenth, eighteenth, twentieth, and twenty-first letters have been published already in the *Philosophical Transactions*; but as nicer observation has furnished several corrections and additions, it is hoped that the republication of them will not give offence; especially as these sheets would be very imperfect without them, and as they will be new to many readers who had no opportunity of seeing them when they made their first appearance.]

The hirundines are a most inoffensive, harmless, entertaining, social, and useful tribe of birds; they touch no fruit in our gardens; delight, all except one species, in attaching themselves to our houses; amuse us with their migrations, songs, and marvellous agility; and clear our outlets from the annoyances of gnats and other troublesome insects. Some districts in the south seas, near Guiaquil[1], are desolated, it seems, by the infinite swarms of venomous mosquitoes, which fill the air, and render those coasts insupportable. It would be worth inquiring whether any species of hirundines is found in those regions. Whoever contemplates the myriads of insects that sport in the sunbeams of a summer evening in this country, will soon be convinced to what a degree our atmosphere would be choked with them was it not for the friendly interposition of the swallow-tribe.

Many species of birds have their peculiar lice; but the hirundines alone seem to be annoyed with dipterous insects, which infest every species, and are so large, in proportion to themselves, that they must be extremely irksome and injurious to them. These are the *hippoboscae hirundinis*, with narrow subulated wings, abounding in every nest; and are hatched by the warmth of the bird's own body during incubation, and crawl about under its feathers.

A species of them is familiar to horsemen in the south of England under the name of forest-fly; and to some of side-fly, from its running sideways like a crab. It creeps under the tails, and about the

1 See *Ulloa's Travels*.

groins of horses, which, at their first coming out of the north, are rendered half frantic by the tickling sensation; while our own breed little regards them.

The curious Reaumur discovered the large eggs, or rather *pupae*, of these flies as big as the flies themselves, which he hatched in his own bosom. Any person that will take the trouble to examine the old nests of either species of swallows may find in them the black shining cases or skins of the *pupae* of these insects; but for other particulars, too long for this place, we refer the reader to *L'Histoire d'Insectes* of that admirable entomologist. Tom. iv., pl. ii.

House-martin

Letter 16 also to the Hon. Daines Barrington

Selborne, November 20th, 1773

Dear Sir – In obedience to your injunctions, I sit down to give you some account of the housemartin, or martlet; and if my monography of this little domestic and familiar bird should happen to meet with your approbation, I may probably soon extend my inquiries to the rest of the British hirundines – the swallow, the swift, and the bank-martin [sand-martin].

A few house-martins begin to appear about the 16th of April; usually some few days later than the swallow. For some time after they appear the hirundines in general pay no attention to the business of nidification, but play and sport about, either to recruit from the fatigue of their journey, if they do migrate at all, or else that their blood may recover its true tone and texture after it has been so long benumbed by the severities of winter. About the middle of May, if the weather be fine, the martin begins to think in earnest of providing a mansion for its family. The crust or shell of this nest seems to be formed of such dirt or loam as comes most readily to hand, and is tempered and wrought together with little bits of broken straws to render it tough and tenacious. As this bird often builds against a perpendicular wall without any projecting ledge under, it requires its utmost efforts to get the first foundation firmly fixed, so that it may safely carry the superstructure. On this occasion the bird not only clings with its claws, but partly supports itself by strongly inclining its tail against the wall, making that a fulcrum; and thus steadied, it works and plasters the materials into the face of the brick or stone. But then, that this work may not, while it is soft and green, pull itself down by its own weight, the provident architect has prudence and forbearance enough not to advance her work too fast; but by building only in the morning, and by dedicating the rest of the day to food and amusement, gives it sufficient time to dry and harden. About half an inch seems to be a sufficient layer for a day. Thus careful workmen, when they build mud-walls (informed at first perhaps by this little bird), raise but a moderate layer at a time, and then desist, lest the work should become top-heavy, and so be ruined by its own weight. By this method in about ten or twelve days is formed an hemispheric nest with a small aperture towards the top, strong, compact, and warm; and perfectly fitted for all the purposes for which it was intended. But then nothing is more common than for the house-sparrow, as soon as the shell is finished, to seize on it as its own, to eject the owner, and to line it after its own manner.[1]

After so much labour is bestowed in erecting a mansion, as nature seldom works in vain, martins will breed on for several years together in the same nest, where it happens to be well sheltered and secure from the injuries of weather. The shell or crust of the nest is a sort of rustic-work full of knobs and protuberances on the outside; nor is the inside of those that I have examined smoothed with any exactness at all; but is rendered soft and warm, and fit for incubation, by a lining of small straws, grasses, and feathers, and sometimes by a bed of moss interwoven with wool. In this nest they tread, or engender, frequently during the time of building; and the hen lays from three to five white eggs.

At first when the young are hatched, and are in a naked and

1 In the case of some house-martins which had built under the eaves of my own cottage, I removed no less than twelve successive nests of sparrows. While each nest was being removed, the sparrows hopped about close by with building materials in their mouths, looking exceedingly saucy, and with an obviously insolent air of self-assertion. They seemed to say, 'Who's afraid of you? We mean to go on building in spite of you.' In the end, however, we tired them out, and the house-martins returned to undisturbed possession. ED.

helpless condition, the parent birds, with tender assiduity, carry out what comes away from their young. Was it not for this affectionate cleanliness the nestlings would soon be burnt up, and destroyed in so deep and hollow a nest, by their own caustic excrement. In the quadruped creation the same neat precaution is made use of; particularly among dogs and cats, where the dams lick away what proceeds from their young. But in birds there seems to be a particular provision, that the dung of nestlings is enveloped in a tough kind of jelly, and therefore is the easier conveyed off without soiling or daubing. Yet, as nature is cleanly in all her ways, the young perform this office for themselves in a little time by thrusting their tails out at the aperture of their nest. As the young of small birds presently arrive at their ἡλικία, or full growth, they soon become impatient of confinement, and sit all day with their heads out at the orifice, where the dams, by clinging to the nest, supply them with food from morning to night. For a time the young are fed on the wing by their parents; but the feat is done by so quick and almost imperceptible a flight that a person must have attended very exactly to their motions before he would be able to perceive it. As soon as the young are able to shift for themselves, the dams immediately turn their thoughts to the business of a second brood; while the first flight, shaken off and rejected by their nurses, congregate in great flocks, and are the birds that are seen clustering and hovering on sunny mornings and evenings round towers and steeples, and on the roofs of churches and houses. These congregatings usually begin to take place about the first week in August; and therefore we may conclude that by that time the first flight is pretty well over. The young of this species do not quit their abodes all together; but the more forward birds get abroad some days before the rest. These approaching the eaves of buildings, and playing about before them, make people think that several old ones attend one nest. They are often capricious in fixing on a nesting-place, beginning many edifices and leaving them unfinished;[2] but when once a nest is completed in a sheltered place, it serves for several seasons. Those which breed in a ready-finished house get the start in hatching of those that build new by ten days or a fortnight. These industrious artificers are at their labours in the long days before four in the morning. When they fix their materials they plaster them on with their chins, moving their heads with a

2 This is usually due either to hostile demonstrations on the part of sparrows or to difficulty in making the walls hang together in the particular situation. ED.

quick vibratory motion. They dip and wash as they fly sometimes in very hot weather, but not so frequently as swallows. It has been observed that martins usually build to a north-east or north-west aspect, that the heat of the sun may not crack and destroy their nests; but instances are also remembered where they bred for many years in vast abundance in a hot stifled inn-yard against a wall facing to the south.

Birds in general are wise in their choice of situation; but in this neighbourhood every summer is seen a strong proof to the contrary at an house without eaves in an exposed district, where some martins build year by year in the corners of the windows. But, as the corners of these windows (which face to the south-east and south-west) are too shallow, the nests are washed down every hard rain; and yet these birds drudge on to no purpose from summer to summer, without changing their aspect or house. It is a piteous sight to see them labouring when half their nest is washed away and bringing dirt . . . '*generis lapsi sarcire ruinas*'. Thus is instinct a most wonderful unequal faculty; in some instances so much above reason, in other respects so far below it! Martins love to frequent towns, especially if there are great lakes and rivers at hand; nay they even affect the close air of London. And I have not only seen them nesting in the Borough, but even in the Strand and Fleet Street; but then it was obvious from the dinginess of their aspect that their feathers partook of the filth of that sooty atmosphere. Martins are by far the least agile of the four species; their wings and tails are short, and therefore they are not capable of such surprising turns, and quick and glancing evolutions as the swallow. Accordingly they make use of a placid easy motion in a middle region of the air, seldom mounting to any great height, and never sweeping long together over the surface of the ground or water. They do not wander far for food, but affect sheltered districts, over some lake, or under some hanging wood, or in some hollow vale, especially in windy weather. They breed the latest of all the swallow kind: in 1772 they had nestlings on to October 21st, and are never without unfledged young as late as Michaelmas.

As the summer declines the congregating flocks increase in numbers daily by the constant accession of the second broods; till at last they swarm in myriads upon myriads round the villages on the Thames, darkening the face of the sky as they frequent the aits [eyots] of that river, where they roost. They retire, the bulk of them I mean, in vast flocks together about the beginning of October; but have appeared of late years in a considerable flight in this

Gilbert White's Oak

neighbourhood, for one day or two, as late as November the 3rd and 6th, after they were supposed to have been gone for more than a fortnight. They therefore withdraw with us the latest of any species. Unless these birds are very short-lived indeed, or unless they do not return to the district where they are bred, they must undergo vast devastations somehow, and somewhere; for the birds that return yearly bear no manner of proportion to the birds that retire.

House-martins are distinguished from their congeners by having their legs covered with soft downy feathers down to their toes. They are no songsters; but twitter in a pretty inward soft manner in their nests.[3] During the time of breeding they are often greatly molested with fleas. I am, &c.

3 They also call to their young on the wing as they approach the nests to feed them. ED.

Beeding

Letter 17 also to the Hon. Daines Barrington

Ringmer, near Lewes, December 9th, 1773

Dear Sir – I received your last favour just as I was setting out for this place; and am pleased to find that my monography met with your approbation. My remarks are the result of many years' observation; and are I trust true in the whole, though I do not pretend to say that they are perfectly void of mistake, or that a more nice observer might not make many additions, since subjects of this kind are inexhaustible.

If you think my letter worthy the notice of your respectable society, you are at liberty to lay it before them; and they will consider it, I hope, as it was intended, as an humble attempt to promote a more minute inquiry into natural history; into the life and conversation of animals. Perhaps, hereafter, I may be induced to take the house-swallow under consideration; and from that proceed to the rest of the British hirundines.

Though I have now travelled the Sussex Downs upwards of thirty years, yet I still investigate that chain of majestic mountains[1] with fresh admiration year by year; and I think I see new beauties every time I traverse it. This range, which runs from Chichester eastward as far as East Bourn [Eastbourne], is about sixty miles in length, and is called the South Downs, properly speaking, only round Lewes. As

1 There is no passage in White more redolent of the eighteenth-century manner of regarding natural scenery than this. ED.

you pass along, you command a noble view of the wild,[2] or weald, on one hand, and the broad downs and sea on the other. Mr Ray used to visit a family[3] just at the foot of these hills, and was so ravished with the prospect from Plumpton Plain, near Lewes, that he mentions those scapes[4] in his 'Wisdom of God in the Works of the Creation' with the utmost satisfaction, and thinks them equal to anything he had seen in the finest parts of Europe.

For my own part, I think there is somewhat peculiarly sweet and amusing in the shapely figured aspect of chalk-hills in preference to those of stone, which are rugged, broken, abrupt, and shapeless.[5]

Perhaps I may be singular in my opinion, and not so happy as to convey to you the same idea; but I never contemplate these mountains without thinking I perceive somewhat analogous to growth in their gentle swellings and smooth fungus-like protuberances, their fluted sides, and regular hollows and slopes, that carry at once the air of vegetative dilatation and expansion . . . Or was there ever a time when these immense masses of calcareous matter were thrown into fermentation by some adventitious moisture; were raised and leavened into such shapes by some plastic power; and so made to swell and heave their broad backs into the sky so much above the less animated clay of the wild [weald] below? [6]

By what I can guess from the admeasurements of the hills that have been taken round my house, I should suppose that these hills sur-

2 The 'wild' is of course superfine English for the weald. The word weald is a good old term for a wooded district, and is allied to the German Wald. The whole of the weald of Surrey and Sussex was once covered by a dense oak forest; even now it is very thickly wooded. Both in this instance, and at Monkton Weald in Dorsetshire, the people preserve the true pronunciation, though White here ignorantly writes 'wild' and the ordnance surveyors write 'Monkton Wyld'. As a rule, in such cases, the popular form is the correct one, while 'educated' people, striving to be more correct, distort or lose sight of the true etymology. ED.

3 Mr Courthope of Danny.

4 Views. This is a rare example of the separate use of the word, familiar to us all in composition in 'landscape' and 'seascape'. ED.

5 Here again we get the eighteenth-century notion that rugged and rocky scenery is 'shapeless', and therefore, ugly. What that age specially admired was smiling cultivation; wild mountainous districts it regarded as repellent and terrifying. ED.

6 We now know that these shapes are due to the slow denuding action of rain-water, which gradually melts away the surface of the chalk beneath the thin layer of turf which covers it. The weald clay underlies the chalk, a thick mass of which once spread over it from North Downs to South

mount the wild at an average of about the rate of five hundred feet.

One thing is very remarkable as to the sheep: from the westward till you get to the river Adur all the flocks have horns, and smooth white faces, and white legs, and a hornless sheep is rarely to be seen; but as soon as you pass that river eastward, and mount Beeding Hill, all the flocks at once become hornless, or as they call them, poll-sheep; and have, moreover, black faces with a white tuft of wool on their foreheads, and speckled and spotted legs, so that you would think that the flocks of Laban were pasturing on one side of the stream, and the variegated breed of his son-in-law Jacob were cantoned along on the other. And this diversity holds good respectively on each side from the valley of Bramber and Beeding to the eastward, and westward all the whole length of the downs. If you talk with the shepherds on this subject, they tell you that the case has been so from time immemorial; and smile at your simplicity if you ask them whether the situation of these two different breeds might not be reversed? However, an intelligent friend of mine near Chichester is determined to try the experiment; and has this autumn, at the hazard of being laughed at, introduced a parcel of black-faced hornless rams among his horned western ewes. The black-faced poll-sheep have the shortest legs and the finest wool.

As I had hardly ever before travelled these downs at so late a season of the year, I was determined to keep as sharp a look-out as possible so near the southern coast, with respect to the summer short-winged birds of passage. We make great inquiries concerning the withdrawing of the swallow-kind, without examining enough into the causes why this tribe is never to be seen in winter; for, *entre nous*, the disappearing of the latter is more marvellous than that of the former, and much more unaccountable. The hirundines, if they please, are certainly capable of migration, and yet no doubt are often found in a torpid state; but redstarts, nightingales, white-throats, black-caps, &c. &c., are very ill provided for long flights; have never been once found, as I ever heard of,[7] in a torpid state; and yet can never be supposed, in such troops, from year to year to dodge and elude the eyes of the curious and inquisitive, which from day to day

---

Downs. The central portion of this sheet of chalk has long since been removed by denudation; the harder mass to the north and south still overlies the clay, but is itself in process of receding slowly. ED.

7 A good example of a phrase, in itself excellent English, but now, by mere disuse, degraded into a vulgarism. ED.

discern the other small birds that are known to abide our winters. But, notwithstanding all my care, I saw nothing like a summer bird of passage; and, what is more strange, not one wheatear, though they abound so in the autumn as to be a considerable perquisite to the shepherds that take them; and though many are to be seen to my knowledge all the winter through in many parts of the south of England. The most intelligent shepherds tell me that some few of these birds appear on the downs in March, and then withdraw to breed probably in warrens and stone-quarries; now and then a nest is ploughed up in a fallow on the downs under a furrow, but it is thought a rarity. At the time of wheat-harvest they begin to be taken in great numbers; are sent for sale in vast quantities to Brighthelmstone [Brighton] and Tunbridge; and appear at the tables of all the gentry that entertain with any degree of elegance. About Michaelmas they retire and are seen no more till March. Though these birds are, when in season, in great plenty on the south downs round Lewes, yet at East Bourn [Eastbourne], which is the eastern extremity of those downs, they abound much more.

One thing is very remarkable, that though in the height of the season so many hundreds of dozens are taken, yet they never are seen to flock; and it is a rare thing to see more than three or four at a time; so that there must be a perpetual flitting and constant progressive succession. It does not appear that any wheatears are taken to the westward of Houghton Bridge, which stands on the river Arun.

I did not fail to look particularly after my new migration of ring-ousels; and to take notice whether they continued on the downs in this season of the year; as I had formerly remarked them in the month of October all the way from Chichester to Lewes wherever there were any shrubs and covert: but not one bird of this sort came within my observation. I only saw a few larks and whinchats, some rooks, and several kites and buzzards.

About Midsummer a flight of cross-bills comes to the pine-groves about this house, but never makes any long stay.

The old tortoise, that I have mentioned in a former letter, still continues in this garden; and retired under ground about the twentieth of November, and came out again for one day on the thirtieth: it lies now buried in a wet swampy border under a wall facing to the south, and is enveloped at present in mud and mire.[8]

8 The shell of this historical tortoise, presented by White's niece to the British Museum, is now in their natural history collection at South Kensington. ED.

Hucker's Lane

There is a large rookery round this house, the inhabitants of which seem to get their livelihood very easily; for they spend the greatest part of the day on their nest-trees when the weather is mild. These rooks retire every evening all the winter from this rookery, where they only call by the way, as they are going to roost in deep woods: at the dawn of day they always revisit their nest-trees, and are preceded a few minutes by a flight of daws, that act, as it were, as their harbingers.

I am, &c.

Swallow

Letter 18 also to the Hon. Daines Barrington

Selborne, January 29th, 1774

Dear Sir – The house-swallow, or chimney-swallow, is undoubtedly the first comer of all the British hirundines; and appears in general on or about the thirteenth of April, as I have remarked from many years' observation.[1] Not but now and then a straggler is seen much earlier; and, in particular, when I was a boy I observed a swallow for a whole day together on a sunny warm Shrove Tuesday; whichday could not fall out later than the middle of March, and often happened early in February.

It is worth remarking that these birds are seen first about lakes and mill-ponds; and it is also very particular, that if these early visitors happen to find frost and snow, as was the case of the two

1 Later observers are almost unanimous in noting that the sand-martin precedes the chimney-swallow by from seven to eleven days. This is also my own experience. On Hind Head, sand-martins flit round the houses, catching flies, till the house-martins return to their nests; but after the house-martins have arrived, the sand-martins abandon the neighbourhood of the houses, and hawk only on the open moors. ED.

dreadful springs of 1770 and 1771, they immediately withdraw for a time. A circumstance this much more in favour of hiding than migration; since it is much more probable that a bird should retire to its hybernaculum just at hand, than return for a week or two only to warmer latitudes.[2]

The swallow, though called the chimney-swallow, by no means builds altogether in chimneys, but often within barns and outhouses against the rafters; and so she did in Virgil's time:

... Ante
Garrula quam tignis nidos suspendat hirundo.

In Sweden she builds in barns, and is called *ladu swala*, the barn swallow. Besides, in the warmer parts of Europe there are no chimneys to houses, except they are English-built: in these countries she constructs her nest in porches, and gateways, and galleries, and open halls.[3]

Here and there a bird may affect some odd, peculiar place; as we have known a swallow build down the shaft of an old well, through which chalk had been formerly drawn up for the purpose of manure: but in general with us this *hirundo* breeds in chimneys; and loves to haunt those stacks where there is a constant fire, no doubt for the sake of warmth. Not that it can subsist in the immediate shaft where there is a fire; but prefers one adjoining to that of the kitchen, and disregards the perpetual smoke of that tunnel, as I have often observed with some degree of wonder.

Five or six or more feet down the chimney does this little bird begin to form her nest about the middle of May, which consists, like that of the house-martin, of a crust or shell composed of dirt or mud, mixed with short pieces of straw to render it tough and permanent; with this difference, that whereas the shell of the martin is nearly hemispheric, that of the swallow is open at the top, and like half a deep dish: this nest is lined with fine grasses, and feathers, which are often collected as they float in the air.

2 Once more the same old pitfall. It is most likely that the birds in these cases were killed by the cold. ED.

3 It is a curious fact that at the present day all the places in which the chimney-swallow builds are of artificial human origin. Hence it seems probable that before the epoch of house-building, the swallow must have bred only in caverns or on accidental cliffs. The enormous growth of human building must, therefore, have admitted of an immense extension of the swallow species. ED.

Wonderful is the address which this adroit bird shows all day long in ascending and descending with security through so narrow a pass. When hovering over the mouth of the funnel, the vibrations of her wings acting on the confined air occasion a rumbling like thunder. It is not improbable that the dam submits to this inconvenient situation so low in the shaft, in order to secure her broods from rapacious birds, and particularly from owls, which frequently fall down chimneys, perhaps in attempting to get at these nestlings.

The swallow lays from four to six white eggs, dotted with red specks; and brings out her first brood about the last week in June, or the first week in July. The progressive method by which the young are introduced into life is very amusing; first, they emerge from the shaft with difficulty enough, and often fall down into the rooms below: for a day or so they are fed on the chimney-top, and then are conducted to the dead leafless bough of some tree, where, sitting in a row, they are attended with great assiduity, and may then be called *perchers*. In a day or two more they become *flyers*, but are still unable to take their own food; therefore they play about near the place where the dams are hawking for flies; and when a mouthful is collected, at a certain signal given, the dam and the nestling advance, rising towards each other, and meeting at an angle; the young one all the while uttering such a little quick note of gratitude and complacency, that a person must have paid very little regard to the wonders of Nature that has not often remarked this feat.[4]

The dam betakes herself immediately to the business of a second brood as soon as she is disengaged from her first, which at once associates with the first broods of house-martins, and with them congregates, clustering on sunny roofs, towers, and trees. This hirundo brings out her second brood towards the middle and end of August.

All the summer long is the swallow a most instructive pattern of unwearied industry and affection; for, from morning to night, while there is a family to be supported, she spends the whole day in skimming close to the ground, and exerting the most sudden turns and quick evolutions. Avenues, and long walks under hedges, and pasture-fields, and mown meadows where cattle graze, are her delight, especially if there are trees interspersed; because in such spots insects most abound. When a fly is taken a smart snap from her bill is heard, resembling the noise at the shutting of a watch-

4 It needs so quick an eye to observe this habit, however, that many people accustomed to note facts of such an order may easily overlook it. ED.

case; but the motion of the mandibles is too quick for the eye.

The swallow, probably the male bird, is the *excubitor* to house martins and other little birds, announcing the approach of birds of prey. For as soon as a hawk appears, with a shrill alarming note he calls all the swallows and martins about him, who pursue in a body, and buffet and strike their enemy till they have driven him from the village, darting down from above on his back, and rising in a perpendicular line in perfect security. This bird also will sound the alarm, and strike at cats when they climb on the roofs of houses, or otherwise approach the nests. Each species of hirundo drinks as it flies along, sipping the surface of the water; but the swallow alone, in general, washes on the wing, by dropping into a pool for many times together: in very hot weather house-martins and bank-martins dip and wash a little.

The swallow is a delicate songster, and in soft sunny weather sings both perching and flying; on trees in a kind of concert, and on chimney-tops: is also a bold flyer, ranging to distant downs and commons even in windy weather, which the other species seem much to dislike; nay, even frequenting exposed sea-port towns, and making little excursions over the salt water. Horsemen on wide downs are often closely attended by a little party of swallows for miles together, which plays before and behind them, sweeping around them, and collecting all the skulking insects that are roused by the trampling of the horses' feet: when the wind blows hard, without this expedient, they are often forced to settle to pick up their lurking prey.

This species feeds much on little *Coleoptera*, as well as on gnats and flies; and often settles on dug ground, or paths, for gravels to grind and digest its food. Before they depart, for some weeks, to a bird, they forsake houses and chimneys, and roost in trees; and usually withdraw about the beginning of October, though some few stragglers may appear on at times till the first week in November.

Some few pairs haunt the new and open streets of London next the fields, but do not enter, like the house-martin, the close and crowded parts of the city.

Both male and female are distinguished from their congeners by the length and forkedness of their tails.[5] They are undoubtedly the

5 The tail, however, is much more forked and much longer in the male than in the female. The difference thus noted is probably ornamental, and is doubtless due to selection of the handsomer partners by the hen birds. ED.

most nimble of all the species: and when the male pursues the female in amorous chase, they then go beyond their usual speed and exert a rapidity almost too quick for the eye to follow.

After this circumstantial detail of the life and discerning στοργὴ of the swallow, I shall add, for your further amusement, an anecdote or two not much in favour of her sagacity:

A certain swallow built for two years together on the handles of a pair of garden-shears that were stuck up against the boards in an out-house, and therefore must have her nest spoiled whenever that implement was wanted; and, what is stranger still, another bird of the same species built its nest on the wings and body of an owl that happened by accident to hang dead and dry from the rafter of a barn. This owl, with the nest on its wings, and with eggs in the nest, was brought as a curiosity worthy the most elegant private museum in Great Britain. The owner, struck with the oddity of the sight, furnished the bringer with a large shell, or conch, desiring him to fix it just where the owl hung: the person did as he was ordered, and the following year a pair, probably the same pair, built their nest in the conch, and laid their eggs.

The owl and the conch make a strange grotesque appearance, and are not the least curious specimens in that wonderful collection of art and nature.[6]

Thus is instinct in animals, taken the least out of its way, an undistinguishing, limited faculty, and blind to every circumstance that does not immediately respect self-preservation, or lead at once to the propagation or support of their species.

I am,

With all respect, &c. &c.

6 Sir Ashton Lever's *Musaeum*.

Letter 19 also to the Hon. Daines Barrington

Selborne, February 14th, 1774

Dear Sir – I received your favour of the eighth, and am pleased to find that you read my little history of the swallow with your usual candour; nor was I the less pleased to find that you made objections where you saw reason.

As to the quotations, it is difficult to say precisely which species of hirundo Virgil might intend in the lines in question, since the ancients did not attend to specific differences like modern naturalists: yet somewhat may be gathered, enough to incline me to suppose that in the two passages quoted the poet had his eye on the swallow.[1]

In the first place the epithet *garrula* suits the swallow well, who is a great songster, and not the martin, who is rather a mute bird; and when it sings is so inward as scarce to be heard. Besides, if *tignum* in that place signifies a rafter rather than a beam, as it seems to me to do, then I think it must be the swallow that is alluded to, and not the martin, since the former does frequently build within the roof

1 This dilettante question of the exact meaning of a classical passage is very much in Daines Barrington's amateurish manner, and very little in Gilbert White's. ED.

against the rafters; while the latter always, as far as I have been able to observe, builds without the roof against eaves and cornices.

As to the simile, too much stress must not be laid on it; yet the epithet *nigra* speaks plainly in favour of the swallow, whose back and wings are very black; while the rump of the martin is milk white, its back and wings blue, and all its under part white as snow. Nor can the clumsy motions (comparatively clumsy) of the martin well represent the sudden and artful evolutions and quick turns which Juturna gave to her brother's chariot, so as to elude the eager pursuit of the enraged Aeneas. The verb *sonat* also seems to imply a bird that is somewhat loquacious.[2]

We have had a very wet autumn and winter, so as to raise the springs to a pitch beyond anything since 1764, which was a remarkable year for floods and high waters. The land-springs which we call lavants,[3] break out much on the downs of Sussex, Hampshire, and Wiltshire. The country people say when the *lavants* rise corn will always be dear; meaning that when the earth is so glutted with water as to send forth springs on the downs and uplands, that the corn-vales must be drowned; and so it has proved for these ten or eleven years past. For land-springs have never obtained more since the memory of man than during that period; nor has there been known a greater scarcity of all sorts of grain, considering the great improvements of modern husbandry. Such a run of wet seasons a century or two ago would, I am persuaded, have occasioned a famine. Therefore pamphlets and newspaper-letters, that talk of combinations, tend to inflame and mislead; since we must not expect plenty till Providence sends us more favourable seasons.

The wheat of last year, all round this district, and in the county of Rutland, and elsewhere, yields remarkably bad; and our wheat on the ground, by the continual late sudden vicissitudes from fierce frost to pouring rains, looks poorly; and the turnips rot very fast.

I am, &c.

2 Nigra *velut magnas domini cum divitis aedes*
Pervolat, et pennis alta atria lustrat hirundo,
Pabula parva legens, nidisque loquacibus escas:
Et nunc porticibus vacuis, nunc humida circum
Stagna sonat.

3 Intermittent springs which burst forth only in very rainy seasons. ED.

Sand-martin

Letter 20 also to the Hon. Daines Barrington

Selborne, February 26th, 1774

Dear Sir – The sand-martin, or bank-martin, is by much the least of any of the British hirundines, and, as far as we have ever seen, the smallest known hirundo, though Brisson asserts that there is one much smaller, and that is the *hirundo esculenta*.

But it is much to be regretted that it is scarce possible for any observer to be so full and exact as he could wish in reciting the circumstances attending the life and conversation of this little bird, since it is *fera natura*, at least in this part of the kingdom, disclaiming all domestic attachments, and haunting wild heaths and commons where there are large lakes; while the other species, especially the swallow and house-martin, are remarkably gentle and domesticated, and never seem to think themselves safe but under the protection of man.[1]

1 This was doubtless the case in White's time, when the open heaths and uplands of England were but little inhabited; but at the present day, sand-martins are very familiar birds in many parts of southern Britain. It must be borne in mind that the immense number of railway-cuttings and of exposed

There are in this parish, in the sand-pits and banks of the lakes of Woolmer forest, several colonies of these birds and yet they are never seen in the village; nor do they at all frequent the cottages that are scattered about in that wild district. The only instance I ever remember where this species haunts any building is at the town of Bishop's Waltham, in this county, where many sand-martins nestle and breed in the scaffold-holes of the back wall of William of Wykeham's stables; but then this wall stands in a very sequestered and retired enclosure, and faces upon a large and beautiful lake. And indeed this species seems so to delight in large waters, that no instance occurs of their abounding but near vast pools or rivers;[2] and in particular it has been remarked that they swarm in the banks of the Thames in some places below London bridge.

It is curious to observe with what different degrees of architectonic skill Providence has endowed birds of the same genus, and so nearly correspondent in their general mode of life! for while the swallow and the house-martin discover the greatest address in raising and securely fixing crusts or shells of loam as cunabula for their young, the bank-martin terebrates a round and regular hole in the sand or earth, which is serpentine, horizontal, and about two feet deep. At the inner end of this burrow does this bird deposit, in a good degree of safety, her rude nest, consisting of fine grasses and feathers, usually goose-feathers, very inartificially laid together.

Perseverance will accomplish anything, though at first one would be disinclined to believe that this weak bird, with her soft and tender bill and claws, should ever he able to bore the stubborn sandbank without entirely disabling herself; yet with these feeble instruments have I seen a pair of them make great despatch, and could remark how much they had scooped that day by the fresh sand which ran down the bank, and was of a different colour from that which lay loose and bleached in the sun.

In what space of time these little artists are able to mine and finish these cavities I have never been able to discover, for reasons given

---

sandbanks on the better-graded roads must have afforded an immensely larger opening for sand-martin enterprise. The species is therefore in all probability much more numerous in individuals now than formerly, and may be observed in many quite populous places. Immense numbers of these pretty birds, for example, build at present in the railway-cutting near Gomshall Station on the South Eastern line. ED.

2 I do not think this is now correct. Sand-martins abound in many dry districts tolerably remote from lakes or rivers. ED.

above; but it would be a matter worthy of observation, where it falls in the way of any naturalist to make his remarks. This I have often taken notice of, that several holes of different depths are left unfinished at the end of summer. To imagine that these beginnings were intentionally made in order to be in the greater forwardness for next spring is allowing perhaps too much foresight and *rerum prudentia* to a simple bird. May not the cause of these *latebrae* being left unfinished arise from their meeting in those places with strata too harsh, hard, and solid for their purpose, which they relinquish, and go to a fresh spot that works more freely? Or may they not in other places fall in with a soil as much too loose and mouldering, liable to flounder, and threatening to overwhelm them and their labours?

One thing is remarkable – that, after some years, the old holes are forsaken and new ones bored; perhaps because the old habitations grow foul and fetid from long use or because they may so abound with fleas as to become untenantable. This species of swallow, moreover, is strangely annoyed with fleas; and we have seen fleas, bed-fleas (*pulex irritans*),[3] swarming at the mouths of these holes, like bees on the stools of their hives.

The following circumstance should by no means be omitted – that these birds do not make use of their caverns by way of hybernacula, as might be expected; since banks so perforated have been dug out with care in the winter, when nothing was found but empty nests.

The sand-martin arrives much about the same time with the swallow, and lays, as she does, from four to six white eggs. But as this species is cryptogame, carrying on the business of nidification, incubation, and the support of its young in the dark, it would not be so easy to ascertain the time of breeding, were it not for the coming forth of the broods, which appear much about the time, or rather somewhat earlier than those of the swallow. The nestlings are supported in common like those of their congeners, with gnats and other small insects; and sometimes they are fed with *libellulae* (dragonflies) almost as long as themselves. In the last week in June we have seen a row of these sitting on a rail near a great pool as perchers, and so young and helpless, as easily to be taken by hand; but whether the dams ever feed them on the wing, as swallows and house-martins do, we have never yet been able to determine; nor

3 White is here mistaken. The flea which infests the sand-martin is a special species. ED.

do we know whether they pursue and attack birds of prey.

When they happen to breed near hedges and enclosures, they are dispossessed of their breeding-holes by the house-sparrow, which is on the same account a fell adversary to house-martins.

These hirundines are no songsters, but rather mute, making only a little harsh noise when a person approaches their nests. They seem not to be of a sociable turn, never with us congregating with their congeners in the autumn. Undoubtedly they breed a second time, like the house-martin and swallow, and withdraw about Michaelmas.

Though in some particular districts they may happen to abound, yet on the whole, in the south of England at least, is this much the rarest species. For there are few towns or large villages but what abound with house-martins; few churches, towers, or steeples, but what are haunted by some swifts; scarce a hamlet or single cottage chimney that has not its swallow; while the bank-martins, scattered here and there, live a sequestered life among some abrupt sand-hills, and in the banks of some few rivers.[4]

These birds have a peculiar manner of flying; flitting about with odd jerks, and vacillations, not unlike the motions of a butterfly. Doubtless the flight of all hirundines is influenced by, and adapted to, the peculiar sort of insects which furnish their food. Hence it would be worth inquiry to examine what particular genus of insects affords the principal food of each respective species of swallow.

Notwithstanding what has been advanced above, some few sand-martins, I see, haunt the skirts of London, frequenting the dirty pools in Saint George's Fields, and about Whitechapel. The question is where these build, since there are no banks or bold shores in that neighbourhood; perhaps they nestle in the scaffold-holes of some old or new deserted building. They dip and wash as they fly sometimes, like the house-martin and swallow.

Sand-martins differ from their congeners in the diminutiveness of their size, and in their colour, which is what is usually called a mouse-colour. Near Valencia, in Spain, they are taken, says Willughby, and sold in the markets for the table; and are called by the country people, probably from their desultory jerking manner of flight, *Papilion de Montagna*.

4 All this is now changed, and even in White's own immediate area the sandmartin has become extremely common. ED.

Swift

Letter 21 also to the Hon. Daines Barrington

Selborne, September 28th, 1774

Dear Sir—As the swift or black martin is the largest of the British hirundines,[1] so it is undoubtedly the latest comer. For I remember but one instance of its appearing before the last week in April; and in some of our late frosty, harsh springs, it has not been seen till the beginning of May. This species usually arrives in pairs.

The swift, like the sand-martin, is very defective in architecture, making no crust, or shell, for its nest; but forming it of dry grasses and feathers, very rudely and inartificially put together. With all my attention to these birds, I have never been able once to discover one in the act of collecting or carrying in materials; so that I have suspected (since their nests are exactly the same) that they sometimes usurp upon the house-sparrows, and expel them, as sparrows

1 As I have already noted, the swift is now known to belong to an entirely different group of birds from the swallows, being in reality much more closely related to the tropical humming-birds. Its apparent resemblance to the swallow tribe is purely adaptive, and results only from similarity of habits. Mr Alfred Russel Wallace has worked up this question admirably in his *Tropical Nature*. ED.

do the house and sand-martin; well remembering that I have seen them squabbling together at the entrance of their holes, and the sparrows up in arms, and much disconcerted at these intruders. And yet I am assured, by a nice observer in such matters, that they do collect feathers for their nests in Andalusia, and that he has shot them with such materials in their mouths.[2]

Swifts, like sand-martins, carry on the business of nidification quite in the dark, in crannies of castles, and towers, and steeples, and upon the tops of the walls of churches under the roof; and therefore cannot be so narrowly watched as those species that build more openly; but, from what I could ever observe, they begin nesting about the middle of May; and I have remarked, from eggs taken, that they have sat hard by the ninth of June. In general they haunt tall buildings, churches, and steeples, and breed only in such; yet in this village some pairs frequent the lowest and meanest cottages, and educate their young under those thatched roofs. We remember but one instance where they breed out of buildings, and that is in the sides of a deep chalk-pit near the town of Odiham, in this county, where we have seen many pairs entering the crevices, and skimming and squeaking round the precipices.

As I have regarded these amusive birds with no small attention, if I should advance something new and peculiar with respect to them, and different from all other birds, I might perhaps be credited, especially as my assertion is the result of many years' exact observation. The fact that I would advance is, that swifts tread, or copulate, on the wing; and I would wish any nice observer, that is startled at this supposition, to use his own eyes, and I think he will soon be convinced. In another class of animals, viz. the insect, nothing is so common as to see the different species of many genera in conjunction as they fly. The swift is almost continually on the wing; and as it never settles on the ground, on trees, or roofs, would seldom find opportunity for amorous rites, was it not enabled to indulge them in the air. If any person would watch these birds of a fine morning in May, as they are sailing round at a great height from the ground, he would see, every now and then, one drop on the back of another, and both of them sink down together for many fathoms with a loud piercing shriek. This I take to be the juncture when the business of generation is carrying on.

2 It is now known that the swift collects materials for its nest on the wing in the same way as other birds of similar habits. ED.

As the swift eats, drinks, collects materials for its nest, and, as it seems, propagates on the wing, it appears to live more in the air than any other bird, and to perform all functions there save those of sleeping and incubation.

This hirundo differs widely from its congeners in laying invariably but two eggs at a time, which are milk-white, long, and peaked at the small end; whereas the other species lay at each brood from four to six.[3] It is a most alert bird, rising very early, and retiring to roost very late; and is on the wing in the height of summer at least sixteen hours. In the longest days it does not withdraw to rest till a quarter before nine in the evening, being the latest of all day-birds. Just before they retire whole groups of them assemble high in the air, and squeak, and shoot about with wonderful rapidity. But this bird is never so much alive as in sultry thundery weather, when it expresses great alacrity, and calls forth all its powers. In hot mornings, several, getting together in little parties, dash round the steeples and churches, squeaking as they go in a very clamorous manner; these, by nice observers, are supposed to be males serenading their sitting hens; and not without reason, since they seldom squeak till they come close to the walls or eaves, and since those within utter at the same time a little inward note of complacency.

When the hen has sat hard all day, she rushes forth just as it is almost dark, and stretches and relieves her weary limbs, and snatches a scanty meal for a few minutes, and then returns to her duty of incubation. Swifts, when wantonly and cruelly shot while they have young, discover a little lump of insects in their mouths, which they pouch and hold under their tongue. In general they feed in a much higher district than the other species; a proof that gnats and other insects do also abound to a considerable height in the air; they also range to vast distances, since locomotion is no labour to them who are endowed with such wonderful powers of wing. Their powers seem to be in proportion to their levers; and their wings are longer in proportion than those of almost any other bird. When they mute, or ease themselves in flight, they raise their wings, and make them meet over their backs.

3 White correctly observes the many points of difference between swifts and swallows, but the ideas prevalent in his age prevent him from seeing that the differences are fundamental, the resemblances superficial and adaptive only. Since Darwin's time we have learned to take a different view of such questions. ED.

A Selborne Hop Garden

At certain times in the summer I had remarked that swifts were hawking very low for hours together over pools and streams; and could not help inquiring into the object of their pursuit that induced them to descend so much below their usual range. After some trouble I found that they were taking *phryganeae*, *ephemerae*, and *libellulae* (cadew-flies [caddis-flies], may-flies, and dragonflies), that were just emerged out of their aurelia state. I then no longer wondered that they should be so willing to stoop for a prey that afforded them such plentiful and succulent nourishment.

They bring out their young about the middle or latter end of July; but as these never become perchers, nor, that ever I could discern, are fed on the wing by their dams, the coming forth of the young is not so notorious as in the other species.

On the 30th of last June I untiled the eaves of a house where many pairs build, and found in each nest only two squab, naked *pulli*; on the 8th of July I repeated the same inquiry, and found that they had made very little progress towards a fledged state, but were still naked and helpless. From whence we may conclude that birds whose way of life keeps them perpetually on the wing would not be able to quit their nest till the end of the month. Swallows and martins, that have numerous families, are continually feeding them every two or three minutes; while swifts, that have but two young to maintain, are much at their leisure, and do not attend on their nests for hours together.

Sometimes they pursue and strike at hawks that come in their way; but not with that vehemence and fury that swallows express on the same occasion. They are out all day long in wet days, feeding about and disregarding still rain: from whence two things may be gathered; first, that many insects abide high in the air, even in rain; and next, that the feathers of these birds must be well preened to resist so much wet. Windy, and particularly windy weather with heavy showers, they dislike; and on such days withdraw, and are scarce ever seen.

There is a circumstance respecting the colour of swifts, which seems not to be unworthy of our attention. When they arrive in the spring, they are all over of a glossy, dark soot colour, except their chins, which are white; but, by being all day long in the sun and air, they become quite weather-beaten and bleached before they depart, and yet they return glossy again in the spring. Now, if they pursue the sun into lower latitudes, as some suppose, in order to enjoy a perpetual summer, why do they not return bleached? Do they not rather perhaps retire to rest for a season, and at that juncture moult and change their feathers, since all other birds are known to moult soon after the season of breeding?

Swifts are very anomalous in many particulars, dissenting from all their congeners not only in the number of their young, but in breeding but once in a summer; whereas all the other British hirundines breed invariably twice.[4] It is past all doubt that swifts can breed but once, since they withdraw in a short time after the flight of their young, and some time before their congeners bring out their second broods. We may here remark that, as swifts breed but once in a summer, and only two at a time, and the other hirundines twice, the latter, who lay from four to six eggs, increase at an average five times as fast as the former.

But in nothing are swifts more singular than in their early retreat. They retire, as to the main body of them, by the tenth of August,

4 See note on page 208. ED.

and sometimes a few days sooner; and every straggler invariably withdraws by the 20th, while their congeners, all of them, stay till the beginning of October; many of them all through that month and some occasionally to the beginning of November. This early retreat is mysterious and wonderful, since that time is often the sweetest season in the year. But what is more extraordinary, they begin to retire still earlier in the most southerly parts of Andalusia, where they can be in no ways influenced by any defect of heat, or, as one might suppose, failure of food. Are they regulated in their motions with us by a defect of food, or by a propensity to moulting, or by a disposition to rest after so rapid a life, or by what? This is one of those incidents in natural history that not only baffles our searches, but almost eludes our guesses.[5]

These hirundines never perch on trees or roofs, and so never congregate with their congeners. They are fearless while haunting their nesting-places, and are not to be scared with a gun; and are often beaten down with poles and cudgels as they stoop to go under the eaves. Swifts are much infested with those pests to the genus called *hippoboscae hirundinis*, and often wriggle and scratch themselves in their flight to get rid of that clinging annoyance.

Swifts are no songsters, and have only one harsh screaming note; yet there are ears to which it is not displeasing, from an agreeable association of ideas, since that note never occurs but in the most lovely summer weather.

They never can settle on the ground but through accident; and when down, can hardly rise, on account of the shortness of their legs and the length of their wings; neither can they walk, but only crawl; but they have a strong grasp with their feet, by which they cling to walls. Their bodies being flat they can enter a very narrow crevice; and where they cannot pass on their bellies they will turn up edgewise.

The particular formation of the foot discriminates the swift from all the British hirundines, and indeed from all other known birds, the *hirundo melba*, or great white-bellied swift of Gibraltar, excepted; for it is so disposed as to carry '*omnes quatuor digitos anticos*'—all its four toes forward; besides, the least toe, which should be the back toe, consists of one bone alone, and the other three only of two apiece—a construction most rare and peculiar, but nicely adapted to the purposes in which their feet are employed. This, and some peculiarities attending the nostrils and under

5 John Antony Scopoli M.D., of Carniola.

mandible,[6] have induced a discerning naturalist to suppose that this species might constitute a genus *per se*.

In London a party of swifts frequents the Tower, playing and feeding over the river just below the bridge; others haunt some of the churches of the Borough, next the fields, but do not venture, like the house-martin, into the close crowded part of the town.

The Swedes have bestowed a very pertinent name on this swallow, calling it 'ring swala', from the perpetual rings or circles that it takes round the scene of its nidification.

Swifts feed on *coleoptera*, or small beetles with hard cases over their wings, as well as on the softer insects; but it does not appear how they can procure gravel to grind their food, as swallows do, since they never settle on the ground. Young ones, overrun with *hippoboscae*, are sometimes found, under their nests, fallen to the ground, the number of vermin rendering their abode insupportable any longer. They frequent in this village several abject cottages; yet a succession still haunts the same unlikely roofs – a good proof this that the same birds return to the same spots. As they must stoop very low to get up under these humble eaves, cats lie in wait, and sometimes catch them on the wing.

On the 5th of July, 1775, I again untiled part of a roof over the nest of a swift. The dam sat in the nest; but so strongly was she affected by a natural σροργὴ for her brood, which she supposed to be in danger, that, regardless of her own safety, she would not stir, but lay sullenly by them, permitting herself to be taken in hand. The squab young we brought down and placed on the grass-plot, where they tumbled about, and were as helpless as a new-born child. While we contemplated their naked bodies, their unwieldy disproportioned abdomina, and their heads, too heavy for their necks to support, we could not but wonder when we reflected that these shiftless beings in a little more than a fortnight would be able to dash through the air almost with the inconceivable swiftness of a meteor; and perhaps in their emigration, must traverse vast continents and oceans as distant as the equator. So soon does Nature advance small birds to their ἡλικία or state of perfection; while the progressive growth of men and large quadrupeds is slow and tedious.

I am, &c.

6 It is really the question of food-supply that regulates their movements. ED.

Missel-thrush

Letter 22 also to the Hon. Daines Barrington

Selborne, September 13th, 1774

Dear Sir—By means of a straight cottage chimney, I had an opportunity this summer of remarking, at my leisure, how swallows ascend and descend through the shaft; but my pleasure in contemplating the address with which this feat was performed to a considerable depth in the chimney was somewhat interrupted by apprehensions lest my eyes might undergo the same fate with those of Tobit.[1]

Perhaps it may be some amusement to you to hear at what times the different species of hirundines arrived this spring in three very distant counties of this kingdom. With us the swallow was seen first on April the 4th, the swift on April the 24th, the bank-martin on April the 12th, and the house-martin not till April the 30th. At South Zele, Devonshire, swallows did not arrive till April the 25th, swifts in plenty on May the 1st, and house martins not till the middle of May. At Blackburn, in Lancashire, swifts were seen April the 28th, swallows April the 29th, house-martins May the 1st. Do

1 Tobit ii: 10.

these different dates, in such distant districts, prove anything for or against migration?

A farmer, near Weyhill, fallows his land with two teams of asses; one of which works till noon, and the other in the afternoon. When these animals have done their work, they are penned all night, like sheep, on the fallow. In the winter they are confined and foddered in a yard, and make plenty of dung.

Linnaeus says that hawks '*paciscuntur inducias cum avibus, quamdiu cuculus cuculat*'; but it appears to me, that during that period, many little birds are taken and destroyed by birds of prey, as may be seen by their feathers left in lanes and under hedges.

The missel-thrush is, while breeding, fierce and pugnacious, driving such birds as approach its nest with great fury to a distance. The Welsh call it 'pen y llwyn', the head or master of the coppice. He suffers no magpie, jay, or blackbird, to enter the garden where he haunts; and is, for the time, a good guard to the new-sown legumens. In general, he is very successful in the defence of his family; but once I observed in my garden that several magpies came determined to storm the nest of a missel-thrush: the dams defended their mansion with great vigour, and fought resolutely *pro aris et focis*; but numbers at last prevailed, they tore the nest to pieces, and swallowed the young alive.

In the season of nidification the wildest birds are comparatively tame. Thus the ring-dove breeds in my fields, though they are continually frequented: and the missel-thrush, though most shy and wild in the autumn and winter, builds in my garden close to a walk where people are passing all day long.

Wall-fruit abounds with me this year; but my grapes, that used to be forward and good, are at present backward beyond all precedent: and this is not the worst of the story; for the same ungenial weather, the same black cold solstice, has injured the more necessary fruits of the earth, and discoloured and blighted our wheat. The crop of hops promises to be very large.

Frequent returns of deafness incommode me sadly, and half disqualify me for a naturalist; for when those fits are upon me, I lose all the pleasing notices and little intimations arising from rural sounds; and May is to me as silent and mute with respect to the notes of birds, &c., as August. My eyesight is, thank God, quick and good; but with respect to the other sense, I am, at times, disabled:

And Wisdom at one entrance quite shut out.

Letter 23 also to the Hon. Daines Barrington

Selborne, June 8th, 1775

Dear Sir — On September 21st, 1741, being then on a visit, and intent on field-diversions, I rose before daybreak: when I came into the enclosures, I found the stubbles and clover-grounds matted all over with a thick coat of cobweb, in the meshes of which a copious and heavy dew hung so plentifully that the whole face of the country seemed, as it were, covered with two or three setting nets drawn one over another. When the dogs attempted to hunt, their eyes were so blinded and hood-winked that they could not proceed, but were obliged to lie down and scrape the incumbrances from their faces with their forefeet, so that, finding my sport interrupted, I returned home musing in my mind on the oddness of the occurrence.

As the morning advanced the sun became bright and warm, and the day turned out one of those most lovely ones which no season but the autumn produces; cloudless, calm, serene, and worthy of the South of France itself.

About nine an appearance very unusual began to demand our attention, a shower of cobwebs falling from very elevated regions, and continuing, without any interruption, till the close of the day. These webs were not single filmy threads, floating in the air in all directions, but perfect flakes or rags; some near an inch broad, and five or six long, which fell with a degree of velocity that showed they were considerably heavier than the atmosphere.

On every side as the observer turned his eyes he might behold a continual succession of fresh flakes falling into his sight, and twinkling like stars as they turned their sides towards the sun.

How far this wonderful shower extended would be difficult to say; but we know that it reached Bradley, Selborne, and Alresford, three places which lie in a sort of a triangle, the shortest of whose sides is about eight miles in extent.

At the second of those places there was a gentleman (for whose veracity and intelligent turn we have the greatest veneration) who observed it the moment he got abroad; but concluded that, as soon as he came upon the hill above his house, where he took his morning rides, he should be higher than this meteor, which he imagined might have been blown, like thistledown from the common above; but, to his great astonishment, when he rode to the most elevated part of the down, three hundred feet above his fields, he found the webs in appearance still as much above him as before; still descending into sight in a constant succession, and twinkling in the sun, so as to draw the attention of the most incurious.

Neither before nor after was any such fall observed; but on this day the flakes hung in the trees and hedges so thick that a diligent person sent out might have gathered baskets full.

The remark that I shall make on these cobweb-like appearances, called gossamer, is, that, strange and superstitious as the notions about them were formerly, nobody in these days doubts but that they are the real production of small spiders, which swarm in the fields in fine weather in autumn, and have a power of shooting out webs from their tails so as to render themselves buoyant, and lighter than air. But why these apterous insects should that day take such a wonderful aerial excursion, and why their webs should at once become so gross and material as to be considerably more weighty than air, and to descend with precipitation, is a matter beyond my skill. If I might be allowed to hazard a supposition, I should imagine that those filmy threads, when first shot, might be entangled in the rising dew, and so drawn up, spiders and all, by a brisk evaporation, into the regions where clouds are formed: and if the spiders have a power of coiling and thickening their webs in the air, as Dr Lister says they have [see his Letters to Mr Ray], then, when they were become heavier than the air, they must fall.

Every day in fine weather, in autumn chiefly, do I see those spiders shooting out their webs and mounting aloft: they will go off from your finger, if you will take them into your hand. Last summer one alighted on my book as I was reading in the parlour: and, running to the top of the page, and shooting out a web, took its departure from thence. But what I most wondered at was, that it went off with considerable velocity in a place where no air was stirring; and I am sure that I did not assist it with my breath. So that these little crawlers seem to have, while mounting, some locomotive power without the use of wings, and to move in the air faster than the air itself.

Bramber

Letter 24 also to the Hon. Daines Barrington

Selborne, August 15th, 1775

Dear Sir—There is a wonderful spirit of sociality in the brute creation, independent of sexual attachment: the congregating of gregarious birds in the winter is a remarkable instance.

Many horses, though quiet with company, will not stay one minute in a field by themselves: the strongest fences cannot restrain them. My neighbour's horse will not only not stay by himself abroad, but he will not bear to be left alone in a strange stable without discovering the utmost impatience and endeavouring to break the rack and manger with his fore feet. He has been known to leap out at a stable window, through which dung was thrown, after company; and yet in other respects is remarkably quiet. Oxen and cows will not fatten by themselves; but will neglect the finest pasture that is not recommended by society. It would be needless to instance in sheep, which constantly flock together.[1]

1 All these animals are clearly the descendants of wild gregarious ancestors, in whom the need for sympathy and society has become organic. Social animals pine in solitude: solitary animals, on the contrary, dislike house-mates. ED.

But this propensity seems not to be confined to animals of the same species; for we know a doe, still alive, that was brought up from a little fawn with a dairy of cows; with them it goes a-field, and with them it returns to the yard. The dogs of the house take no notice of this deer, being used to her; but, if strange dogs come by, a chase ensues; while the master smiles to see his favourite securely leading her pursuers over hedge, or gate, or stile, till she returns to the cows, who, with fierce lowings and menacing horns, drive the assailants quite out of the pasture.

Even great disparity of kind and size does not always prevent social advances and mutual fellowship. For a very intelligent and observant person has assured me that, in the former part of his life, keeping but one horse, he happened also on a time to have but one solitary hen. These two incongruous animals spent much of their time together in a lonely orchard, were they saw no creature but each other. By degrees an apparent regard began to take place between these two sequestered individuals. The fowl would approach the quadruped with notes of complacency, rubbing herself gently against his legs: while the horse would look down with satisfaction, and move with the greatest caution and circumspection, lest he should trample on his diminutive companion. Thus, by mutual good offices, each seemed to console the vacant hours of the other: so that Milton, when he puts the following sentiment into the mouth of Adam, seems to be somewhat mistaken:

> 'Much less can *bird* with *beast*, or fish with fowl,
> So well converse, nor with the ox the ape.'

I am, &c.

Letter 25 also to the Hon. Daines Barrington

Selborne, October 2nd, 1775

Dear Sir – We have two gangs or hordes of gypsies which infest the south and west of England, and come round in their circuit two or three times in the year. One of these tribes calls itself by the noble name of Stanley, of which I have nothing particular to say; but the other is distinguished by an appellative somewhat remarkable. As far as their harsh gibberish can be understood, they seem to say that the name of their clan is Curleople; now the termination of this word is apparently Grecian, and as Mezeray and the gravest historians all agree that these vagrants did certainly migrate from Egypt and the East, two or three centuries ago, and so spread by degrees over Europe, may not this family-name, a little corrupted, be the very name they brought with them from the Levant? It would be matter of some curiosity, could one meet with an intelligent person among them, to inquire whether, in their jargon, they still retain any Greek words; the Greek radicals will appear in hand, foot, head, water, earth, &c. It is possible that amidst their cant and corrupted dialect many mutilated remains of their native language might still be discovered.[1]

With regard to those peculiar people, the gypsies, one thing is very remarkable, and especially as they came from warmer climates; and that is, that while other beggars lodge in barns, stables, and cow houses, these sturdy savages seem to pride themselves in braving the severities of winter, and in living *sub dio* the whole year round. Last September was as wet a month as ever was known; and yet during those deluges did a young gypsy girl lie in the midst of one of our hop-gardens, on the cold ground, with nothing over her but a piece of a blanket extended on a few hazel-rods bent hoop-fashion, and stuck into the earth at each end, in circumstances too trying for a cow in the same condition; yet within this garden there was a large hop-kiln, into the chambers of which she might have retired, had she thought shelter an object worthy her attention.

Europe itself, it seems, cannot set bounds to the rovings of these vagabonds; for Mr Bell, in his return from Peking met a gang of these people on the confines of Tartary, who were endeavouring to penetrate those deserts, and try their fortune in China.[2]

Gypsies are called in French, Bohemiens; in Italian and modern Greek, Zingani.

I am, &c.

1 The gypsy language is now known to be an Indian dialect, probably from the Punjab. The fragments which still survive point to a Ját origin. ED.
2 See Bell's *Travels in China*.

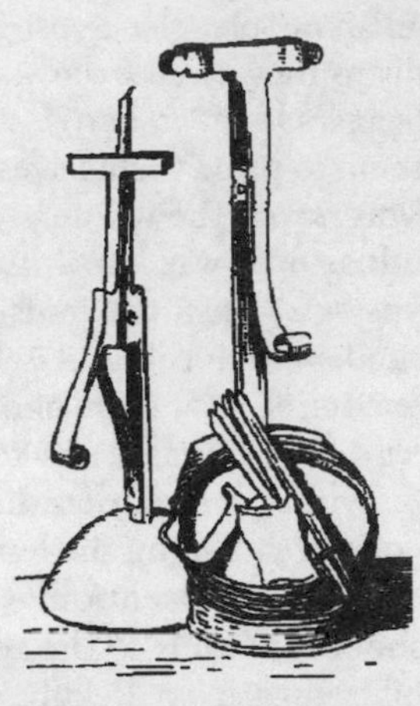

Old rush-lightholders and tinder box

Letter 26 also to the Hon. Daines Barrington

Selborne, November 1st, 1775

Hic . . . taedae pingues, hic plurimus ignis
Semper, et assidua postes fuligine nigri.

Dear Sir — I shall make no apology for troubling you with the detail of a very simple piece of domestic economy being satisfied that you think nothing beneath your attention that tends to utility; the matter alluded to is the use of rushes instead of candles, which I am well aware prevails in many districts besides this; but as I know there are countries also where it does not obtain, and as I have considered the subject with some degree of exactness, I shall proceed in my humble story, and leave you to judge of the expediency.

The proper species of rush for this purpose seems to be the*juncus effusus*, or common soft rush, which is to be found in most moist pastures, by the sides of streams, and under hedges. These rushes are in best condition in the height of summer; but may be gathered, so as to serve the purpose well, quite on to autumn. It would be needless to add that the largest and longest are best. Decayed labourers, women, and children, make it their business to procure and prepare them. As soon as they are cut, they must be flung into water, and kept there, for otherwise they will dry and shrink, and the peel will not run. At first a person would find it no easy matter to divest a rush of its peel or rind, so as to leave one regular, narrow, even rib from top to bottom that may support the pith; but this like other feats, soon becomes familiar even to children; and we have seen an old woman, stone blind, performing this business with great despatch, and seldom failing to strip them with the nicest regularity. When these *junci* are thus far prepared they must lie out on the grass to be bleached, and take the dew for some nights, and afterwards be dried in the sun.

Some address is required in dipping these rushes in scalding fat or grease; but this knack also is to be attained by practice. The careful wife of an industrious Hampshire labourer obtains all her fat for nothing; for she saves the scummings of her bacon-pot for this use: and, if the grease abounds with salt, she causes the salt to precipitate to the bottom, by setting the scummings in a warm oven. Where hogs are not much in use, and especially by the sea-side, the coarser animal-oils will come very cheap. A pound of common grease may be procured for fourpence, and about six pounds of grease will dip a pound of rushes, and one pound of rushes may be bought for one shilling; so that a pound of rushes, medicated and ready for use, will cost three shillings. If men that keep bees will mix a little wax with the grease, it will give it a consistency, and render it more cleanly, and make the rushes burn longer; mutton-suet would have the same effect.

A good rush, which measured in length two feet four inches and a half, being minuted, burnt only three minutes short of an hour; and a rush of still greater length has been known to burn one hour and a quarter.

These rushes give a good clear light. Watch-lights (coated with tallow), it is true, shed a dismal one, 'darkness visible'; but then the wick of those have two ribs of the rind, or peel, to support the pith, while the wick of the dipped rush has but one. The two ribs are intended to impede the progress of the flame and make the candle last.

In a pound of dry rushes, avoirdupois, which I caused to be weighed and numbered, we found upwards of one thousand six hundred individuals. Now suppose each of these burns, one with another, only half an hour, then a poor man will purchase eight hundred hours of light, a time exceeding thirty-three entire days, for three shillings. According to this account each rush, before dipping, costs $1/33$ of a farthing, and $1/11$ afterwards. Thus a poor family will enjoy five and a half hours of comfortable light for a farthing. An experienced old housekeeper assures me that one pound and a half of rushes completely supplies his family the year round, since working people burn no candles in the long days, because they rise and go to bed by daylight.

Little farmers use rushes much in the short days both morning and evening, in the dairy and kitchen; but the very poor, who are always the worst economists, and therefore must continue very poor, buy a halfpenny candle every evening, which in their blowing open rooms, does not burn much more than two hours. Thus they have only two hours' light for their money instead of eleven.

While on the subject of rural economy, it may not be improper to mention a pretty implement of housewifery that we have seen nowhere else; that is, little neat besoms which our foresters make from the stalks of the *polytricum commune*, or great golden maiden hair, which they call silk-wood, and find plenty in the bogs. When this moss is well combed and dressed, and divested of its outer skin, it becomes of a beautiful bright chestnut colour; and, being soft and pliant, is very proper for the dusting of beds, curtains, carpets, hangings, &c. If these besoms were known to the brush-makers in town, it is probable they might come much in use for the purpose above mentioned.[1]

I am, &c.

1 A besom of this sort is to be seen in Sir Ashton Lever's Museum.

On Baker's Hill

Letter 27 also to the Hon. Daines Barrington

Selborne, December 12th, 1775

Dear Sir – We had in this village more than twenty years ago an idiot boy, whom I well remember, who, from a child, showed a strong propensity to bees; they were his food, his amusement, his sole object. And as people of this caste have seldom more than one point in view, so this lad exerted all his few faculties on this one pursuit. In the winter he dozed away his time within his father's house, by the fireside, in a kind of torpid state, seldom departing from the chimney-corner, but in the summer he was all alert, and in quest of his game in the fields, and on sunny banks. Honey-bees, humble-bees, and wasps, were his prey wherever he found them; he had no apprehensions from their stings, but would seize them*nudis manibus*, and at once disarm them of their weapons, and suck their bodies for the sake of their honey-bags. Sometimes he would fill his bosom between his shirt and his skin with a number of these captives, and sometimes would confine them in bottles. He was a very *merops apiaster*, or bee-bird, and very injurious to men that kept bees; for he would slide into their bee-gardens, and, sitting down before the stools, would rap with his finger on the hives, and so take the bees as they came out. He has been known to overturn hives for the sake of honey, of which he was passionately fond. Where metheglin was making[1] he would linger round the tubs and vessels, begging a draught of what he called bee-wine. As he ran about he used to make a humming noise with his lips, resembling the buzzing of bees. This lad was lean and sallow, and of a cadaverous complexion; and, except in his favourite pursuit, in which he was wonderfully adroit, discovered no manner of understanding. Had his capacity been better, and directed to the same object, he had perhaps abated much of our wonder at the feats of a more modern exhibitor of bees; and we may justly say of him now:

> Thou,
> Had thy presiding star propitious shone,
> Should'st Wildman[2] be . . .

When a tall youth he was removed from hence to a distant village where he died, as I understand, before he arrived at manhood.

I am, &c.

1 This is an interesting passage as showing that metheglin was still commonly made in Wessex only a hundred years ago. ED.

2 Thomas Wildman published a *Treatise on the Management of Bees*; with the various methods of cultivating them, both ancient and modern, 4to, 1768.

Letter 28 also to the Hon. Daines Barrington

Selborne, January 8th, 1776

Dear Sir – It is the hardest thing in the world to shake off superstitious prejudices: they are sucked in, as it were, with our mother's milk; and, growing up with us at a time when they take the fastest hold and make the most lasting impressions, become so interwoven into our very constitutions, that the strongest good sense is required to disengage ourselves from them. No wonder, therefore, that the lower people retain them their whole lives through, since their minds are not invigorated by a liberal education, and therefore not enabled to make any efforts adequate to the occasion.

Such a preamble seems to be necessary before we enter on the superstitions of this district, lest we should be suspected of exaggeration in a recital of practices too gross for this enlightened age.

But the people of Tring, in Hertfordshire, would do well to remember, that no longer ago than the year 1751, and within twenty miles of the capital, they seized on two superannuated wretches, crazed with age, and overwhelmed with infirmities, on a suspicion of witchcraft; and, by trying experiments, drowned them in a horse-pond.

In a farmyard near the middle of this village stands, at this day, a row of pollard-ashes, which, by the seams and long cicatrices down their sides, manifestly show that, in former times, they have been cleft asunder. These trees, when young and flexible, were severed and held open by wedges, while ruptured children, stripped naked, were pushed through the apertures, under a persuasion that, by such a process, the poor babes would be cured of their infirmity. As soon as the operation was over, the tree, in the suffering part, was plastered with loam, and carefully swathed up. If the parts coalesced and soldered together, as usually fell out, where the feat was performed with any adroitness at all, the party was cured; but, where the cleft continued to gape, the operation, it was supposed, would prove ineffectual. Having occasion to enlarge my garden not long since, I cut down two or three such trees, one of which did not grow together.

We have several persons now living in the village, who, in their childhood, were supposed to be healed by this superstitious ceremony derived down perhaps from our Saxon ancestors, who practised it before their conversion to Christianity.

At the South corner of the Plestor, or area near the church, there stood, about twenty years ago, a very old grotesque hollow pollard ash, which for ages had been looked on with no small veneration as a shrew-ash. Now a shrew-ash is an ash whose twigs or branches, when gently applied to the limbs of cattle, will immediately relieve the pains which a beast suffers from the running of a shrew-mouse over the part affected; for it is supposed that a shrew-mouse is of so baneful and deleterious a nature, that wherever it creeps over a beast, be it horse, cow, or sheep, the suffering animal is afflicted with cruel anguish, and threatened with the loss of the use of the limb.[1] Against this accident, to which they were continually liable, our provident forefathers always kept a shrew-ash at hand, which,

1 This observation leads up to the modern science of folklore, dealing with a class of facts too often despised in White's time. 'Shrew-struck' horses were frequently cured by dragging the animal through the aperture of a bramble which had grown into the earth at the upper end, as frequently happens. The shrew-ash is a special case of that immolation of the deity of vegetation so fully illustrated in Mr Frazer's *Golden Bough*. ED.

The Plestor

when once medicated, would maintain its virtue for ever. A shrew-ash was made thus: into the body of the tree a deep hole was bored with an auger, and a poor devoted shrew-mouse was thrust in alive, and plugged in, no doubt, with several quaint incantations long since forgotten. As the ceremonies necessary for such a consecration are no longer understood, all succession is at an end, and no such tree is known to subsist in the manor, or hundred.

As to that on the Plestor

The late vicar stubb'd and burnt it,

when he was way-warden, regardless of the remonstrances of the bystanders, who interceded in vain for its preservation, urging its power and efficacy, and alleging that it had been,

Religione patrum multos servata per annos.

I am, &c.

Wood Pond

Letter 29 also to the Hon. Daines Barrington

Selborne, February 7th, 1776

Dear Sir – In heavy fogs, on elevated situations especially, trees are perfect alembics; and no one that has not attended to such matters can imagine how much water one tree will distil in a night's time, by condensing the vapour, which trickles down the twigs and boughs so as to make the ground below quite in a float. In Newton Lane, in October 1775, on a misty day, a particular oak in leaf dropped so fast that the cartway stood in puddles and the ruts ran with water, though the ground in general was dusty.

In some of our smaller islands in the West Indies, if I mistake not, there are no springs or rivers; but the people are supplied with that necessary element, water, merely by the dripping of some large, tall trees, which, standing in the bosom of a mountain, keep their heads constantly enveloped with fogs and clouds, from which they dispense their kindly never-ceasing moisture; and so render those districts habitable by condensation alone.

Trees in leaf have such a vast proportion more of surface than those that are naked, that, in theory, their condensations should greatly exceed those that are stripped of their leaves; but, as the former imbibe also a great quantity of moisture, it is difficult to say which drip most; but this I know, that deciduous trees that are entwined with much ivy seem to distil the greatest quantity. Ivy leaves are smooth, and thick, and cold, and therefore condense very fast; and besides, evergreens imbibe very little. These facts may furnish the intelligent with hints concerning what sorts of trees they should plant round small ponds that they would wish to be perennial; and show them how advantageous some trees are in preference to others.[1]

Trees perspire profusely, condense largely, and check evaporation so much, that woods are always moist; no wonder, therefore, that they contribute much to pools and streams.

That trees are great promoters of lakes and rivers appears from a well-known fact in North America: for, since the woods and forests have been grubbed and cleared, all bodies of water are much diminished; so that some streams, that were very considerable a century ago, will not now drive a common mill.[2] Besides, most woodlands, forests, and chases, with us abound with pools and morasses; no doubt for the reason given above.

To a thinking mind few phenomena are more strange than the state of little ponds on the summits of chalk-hills, many of which are never dry in the most trying droughts of summer. On chalk hills I say, because in many rocky and gravelly soils springs usually break out pretty high on the sides of elevated grounds and mountains: but no person acquainted with chalky districts will allow that they ever saw springs in such a soil but in valleys and bottoms, since the waters of so pervious a stratum as chalk all lie on one dead level, as well-diggers have assured me again and again.

Now we have many such little round ponds in this district; and one in particular on our sheep-down, three hundred feet above my house; which, though never above three feet deep in the middle, and not more than thirty feet in diameter, and containing perhaps not more than two or three hundred hogsheads of water, yet never is known to fail, though it affords drink for three hundred or four hundred sheep, and for at least twenty head of large cattle beside. This pond, it is true, is overhung with two moderate beeches, that, doubtless, at times afford it much supply: but then we have others as

1 *Vide* Kalm's *Travels to North America*.

2 It is now well known that dry districts have had their rainfall largely increased by being planted with trees, while districts once moist have been rendered arid and desert by clearing off the timber. ED.

small, that, without the aid of trees, and in spite of evaporation from sun and wind, and perpetual consumption by cattle, yet constantly maintain a moderate share of water, without overflowing in the wettest seasons, as they would do if supplied by springs. By my journal of May, 1775, it appears that 'the small and even considerable ponds in the vales are now dried up, while the small ponds on the very tops of hills are but little affected'. Can this difference be accounted for from evaporation alone, which certainly is more prevalent in bottoms? or rather have not those elevated pools some unnoticed recruits, which in the night time counterbalance the waste of the day; without which the cattle alone must soon exhaust them? And here it will be necessary to enter more minutely into the cause. Dr Hales, in his *Vegetable Statics*, advances, from experiment, that, 'the moister the earth is the more dew falls on it in a night; and more than a double quantity of dew falls on a surface of water than there does on an equal surface of moist earth'. Hence we see that water, by its coolness, is enabled to assimilate to itself a large quantity of moisture nightly by condensation; and that the air, when loaded with fogs and vapours, and even with copious dews, can alone advance a considerable and never-failing resource. Persons that are much abroad, and travel early and late, such as shepherds, fishermen, &c., can tell what prodigious fogs prevail in the night on elevated downs, even in the hottest parts of summer; and how much the surfaces of things are drenched by those swimming vapours, though, to the senses, all the while, little moisture seems to fall.

I am, &c.

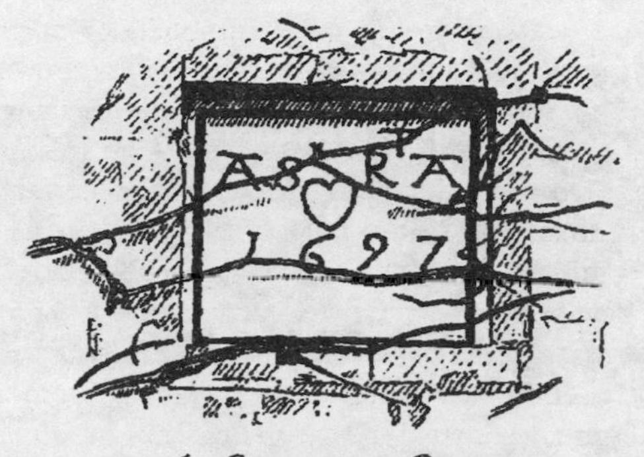

In Gracious Street

Letter 30 also to the Hon. Daines Barrington

Selborne, April 3rd, 1776

Dear Sir – Monsieur Herissant, a French anatomist, seems persuaded that he has discovered the reason why cuckoos do not hatch their own eggs; the impediment, he supposes, arises from the internal structure of their parts, which incapacitates them for incubation. According to this gentleman, the crop, or craw, of a cuckoo does not lie before the sternum at the bottom of the neck, as in the *gallinae*, *columbae*, &c., but immediately behind it, on and over the bowels, so as to make a large protuberance in the belly.[1 and 2]

Induced by this assertion, we procured a cuckoo;[3] and, cutting open the breast-bone, and exposing the intestines to sight, found

1 *Histoire de l'Académie Royale*, 1752.

2 There is nothing whatsoever in the structure of the cuckoo to prevent its hatching its own eggs. The whole question of the curious parasitism of cuckoos and some other birds (such as *Molothrus*) has since been fully elucidated by Darwin. ED.

3 Wherever White speaks thus in the first person plural, we may suspect the letter either of being an added one, or else of being largely cooked up for publication. ED.

The Church from the S·wn 1776

the crop lying as mentioned above. This stomach was large and round, and stuffed hard, like a pincushion, with food, which, upon nice examination, we found to consist of various insects; such as small scarabs, spiders, and dragonflies; the last of which we have seen cuckoos catching on the wing as they were just emerging out of the aurelia state. Among this farrago also were to be seen maggots, and many seeds, which belonged either to gooseberries, currants, cranberries, or some such fruit; so that these birds apparently subsist on insects and fruits; nor was there the least appearance of bones, feathers, or fur, to support the idle notion of their being birds of prey.

The sternum in this bird seemed to us to be remarkably short, between which and the anus lay the crop, or craw, and immediately behind that the bowels against the back-bone.

It must be allowed, as this anatomist observes, that the crop placed just upon the bowels must, especially when full, be in a very uneasy situation during the business of incubation; yet the test will be to examine whether birds that are actually known to sit for certain are not formed in a similar manner. This inquiry I proposed to myself to make with a fern-owl, or goatsucker, as soon as opportunity offered: because, if their formation proves the same, the reason for incapacity in the cuckoo will be allowed to have been taken up somewhat hastily.

Not long after a fern-owl was procured, which, from its habit and shape, we suspected might resemble the cuckoo in its internal construction. Nor were our suspicions ill-grounded; for, upon dissection, the crop, or craw, also lay behind the sternum, immediately on the viscera, between them and the skin of the belly. It was bulky, and stuffed hard with large *phalaenae*, moths of several sorts, and their eggs, which no doubt had been forced out of those insects by the action of swallowing.

Now as it appears that this bird, which is so well known to practise incubation, is formed in a similar manner with cuckoos, Monsieur Herissant's conjecture, that cuckoos are incapable of incubation from the disposition of their intestines, seems to fall to the ground; and we are still at a loss for the cause of that strange and singular peculiarity in the instance of the *cuculus canorus*.

We found the case to be the same with the ring-tail hawk, in respect to formation; and, as far as I can recollect, with the swift; and probably it is so with many more sorts of birds that are not granivorous.

I am, &c.

Letter 31 also to the Hon. Daines Barrington

Selborne, April 29th, 1776

Dear Sir – On August the 4th, 1775, we surprised a large viper, which seemed very heavy and bloated, as it lay in the grass basking in the sun. When we came to cut it up, we found that the abdomen was crowded with young, fifteen in number; the shortest of which measured full seven inches, and were about the size of full-grown earthworms. This little fry issued into the world with the true viper-spirit about them, showing great alertness as soon as disengaged from the belly of the dam: they twisted and wriggled about, and set themselves up, and gaped very wide when touched with a stick, showing manifest tokens of menace and defiance, though as yet they had no manner of fangs that we could find, even with the help of our glasses.

To a thinking mind nothing is more wonderful than that early instinct which impresses young animals with a notion of the situation of their natural weapons, and of using them properly in their own defence, even before those weapons subsist or are formed. Thus a young cock will spar at his adversary before his spurs are grown; and a calf or a lamb will push with their heads before their horns are sprouted. In the same manner did these young adders attempt to bite before their fangs were in being. The dam however was furnished with very formidable ones, which we lifted up (for they fold down when not used) and cut them off with the point of our scissors.

Viper

There was little room to suppose that this brood had ever been in the open air before; and that they were taken in for refuge, at the mouth of the dam, when she perceived that danger was approaching; because then probably we should have found them somewhere in the neck, and not in the abdomen.

Letter 32 also to the Hon. Daines Barrington

Castration has a strange effect: it emasculates both man, beast, and bird, and brings them to a near resemblance of the other sex. Thus eunuchs have smooth unmuscular arms, thighs, and legs; and broad hips, and beardless chins, and squeaking voices. Gelt stags and bucks have hornless heads, like hinds and does. Thus wethers have small horns, like ewes; and oxen large bent horns, and hoarse voices when they low, like cows: for bulls have short straight horns; and though they mutter and grumble in a deep tremendous tone, yet they low in a shrill high key. Capons have small combs and gills, and look pallid about the head like pullets; they also walk without any parade, and hover chickens like hens. Barrow-hogs have also small tusks like sows.

Thus far it is plain that the deprivation of masculine vigour puts a stop to the growth of those parts or appendages that are looked upon as its insignia. But the ingenious Mr Lisle, in his book on husbandry, carries it much farther; for he says that the loss of those insignia alone has sometimes a strange effect on the ability itself: he had a boar so fierce and venereous, that, to prevent mischief, orders were given for his tusks to be broken off. No sooner had the beast suffered this injury than his powers forsook him, and he neglected those females to whom before he was passionately attached, and from whom no fences would restrain him.[1]

1 The question of correlation of organs and functions here touched upon is one of those which received the greatest light from Darwin's investigations: see in particular the chapters on sexual selection in *The Descent of Man*. ED.

Letter 33 also to the Hon. Daines Barrington

Gilbert White's arms

The natural term of an hog's life is little known, and the reason is plain – because it is neither profitable nor convenient to keep that turbulent animal to the full extent of its time: however, my neighbour, a man of substance, who had no occasion to study every little advantage to a nicety, kept an half-bred bantam-sow, who was as thick as she was long, and whose belly swept on the ground, till she was advanced to her seventeenth year, at which period she showed some tokens of age by the decay of her teeth and the decline of her fertility.

For about ten years this prolific mother produced two litters in the year of about ten at a time, and once above twenty at a litter; but, as there were near double the number of pigs to that of teats many died. From long experience in the world this female was grown very sagacious and artful. When she found occasion to converse with a boar she used to open all the intervening gates, and march, by herself, up to a distant farm where one was kept; and when her purpose was served would return by the same means. At the age of about fifteen her litters began to be reduced to four or five; and such a litter she exhibited when in her fatting-pen. She proved, when fat, good bacon, juicy, and tender; the rind, or sward, was remarkably thin. At a moderate computation she was allowed to have been the fruitful parent of three hundred pigs: a prodigious instance of fecundity in so large a quadruped! She was killed in spring 1775.

I am, &c.

Letter 34 also to the Hon. Daines Barrington

Selborne, May 9th, 1776

... admorunt ubera tigres.

Dear Sir – We have remarked in a former letter[1] how much incongruous animals, in a lonely state, may be attached to each other from a spirit of sociality; in this it may not be amiss to recount a different motive which has been known to create as strange a fondness.

My friend had a little helpless leveret brought to him, which the servants fed with milk in a spoon, and about the same time his cat kittened and the young were dispatched and buried. The hare was soon lost, and supposed to be gone the way of most foundlings, to be killed by some dog or cat. However, in about a fortnight, as the master was sitting in his garden in the dusk of the evening, he observed his cat, with tail erect, trotting towards him, and calling with little short inward notes of complacency, such as they use towards their kittens, and something gamboling after, which proved to be the leveret that the cat had supported with her milk, and continued to support with great affection.

1 Letter 24.

Thus was a graminivorous animal nurtured by a carnivorous and predaceous one!

Why so cruel and sanguinary a beast as a cat, of the ferocious genus of *Felis*, the *murium leo*, as Linnaeus calls it, should be affected with any tenderness towards an animal which is its natural prey, is not so easy to determine.

This strange affection probably was occasioned by that desiderium, those tender maternal feelings, which the loss of her kittens had awakened in her breast; and by the complacency and ease she derived to herself from the procuring her teats to be drawn, which were too much distended with milk, till, from habit, she became as much delighted with this foundling as if it had been her real offspring.

This incident is no bad solution of that strange circumstance which grave historians as well as the poets assert, of exposed children being sometimes nurtured by female wild beasts that probably had lost their young. For it is not one whit more marvellous that Romulus and Remus, in their infant state, should be nursed by a she-wolf, than that a poor little sucking leveret should be fostered and cherished by a bloody grimalkin.

> *. . . viridi foetam Mavortis in antro*
> *Procubuisse lupam: geminos huic ubera circum*
> *Ludere pendentes pueros, et lambere matrem*
> *Impavidos: illam tereti cervice reflexam*
> *Mulcere alternos, et corpora fingere lingua.*

The Church from the N·in 1776

Letter 35 also to the Hon. Daines Barrington

Selborne, May 20th, 1777

Dear Sir – Lands that are subject to frequent inundations are always poor; and probably the reason may be because the worms are drowned. The most insignificant insects and reptiles are of much more consequence, and have much more influence in the economy of Nature, than the incurious are aware of; and are mighty in their effect, from their minuteness, which renders them less an object of attention: and from their numbers and fecundity. Earthworms, though in appearance a small and despicable link in the chain of Nature, yet, if lost, would make a lamentable chasm. For to say nothing of half the birds, and some quadrupeds which are almost entirely supported by them, worms seem to be the great promoters of vegetation, which would proceed but lamely without them, by boring, perforating, and loosening the soil, and rendering it pervious to rains and the fibres of plants, by drawing straws and stalks of leaves and twigs into it; and, most of all, by throwing up such infinite numbers of lumps of earth called worm-casts, which, being their excrement, is a fine manure for grain and grass. Worms probably provide new soil for hills and slopes where the rain washes the earth away; and they affect slopes, probably to avoid being flooded.[1] Gardeners and farmers express their detestation of worms; the former because they render their walks unsightly, and make them much work; and the latter because, as they think, worms eat their green corn. But these men would find that the earth without worms would soon become cold, hard-bound, and void of fermentation, and consequently sterile; and, besides, in favour of worms, it should be hinted that green corn, plants, and flowers, are not so much injured by them as by many species of *coleoptera* (scarabs), and *tipulae* (long-legs) in their larva, or grub-state; and by unnoticed myriads of small shell-less snails, called slugs, which silently and imperceptibly make amazing havoc in the field and garden.[2]

These hints we think proper to throw out in order to set the inquisitive and discerning to work.

A good monography of worms would afford much entertainment and information at the same time, and would open a large and new field in natural history.[3] Worms work most in the spring; but by no means lie torpid in the dead months: are out every mild night in the winter, as any person may be convinced that will take the pains to examine his grass-plots with a candle; are hermaphrodites, and much addicted to venery, and consequently very prolific.

I am, &c.

1 This very interesting passage gives in brief; but without any full detail of experiments or observations, the main principles afterwards so fully worked out by Darwin in his wonderful treatise on vegetable mould and earthworms. Oddly enough, Darwin, by a rare slip of memory in so candid and accurate a writer, does not allude in his treatise to this passage, from which he must almost certainly have derived the first impetus towards his long and patient investigation of the subject. ED.

2 Farmer Young, of Norton Farm, says, that this spring (1777) about four acres of his wheat in one field was entirely destroyed by slugs, which swarmed on the blades of corn, and devoured it as fast as it sprang.

3 The 'monography' here desired has since been amply supplied by Darwin. ED.

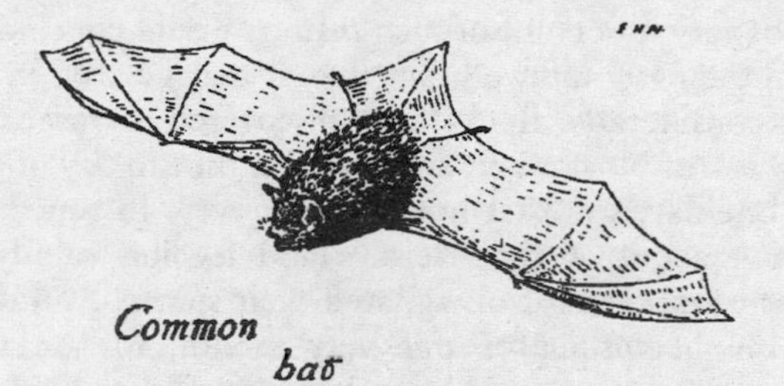

Letter 36 also to the Hon. Daines Barrington[1]

Selborne, November 22nd, 1777

Dear Sir – You cannot but remember that the 26th and 27th of last March were very hot days – so sultry that everybody complained and were restless under those sensations to which they had not been reconciled by gradual approaches.

This sudden summer-like heat was attended by many summer coincidences; for on those two days the thermometer rose to sixty-six in the shade; many species of insects revived and came forth; some bees swarmed in this neighbourhood; the old tortoise, near Lewes, in Sussex, awakened and came forth out of its dormitory; and, what is most to my present purpose, many house-swallows appeared and were very alert in many places, and particularly at Chobham, in Surrey.

But as that short warm period was succeeded as well as preceded by harsh severe weather, with frequent frosts and ice, and cutting winds, the insects withdrew, the tortoise retired again into the ground, and the swallows were seen no more until the 10th of April, when, the rigour of the spring abating, a softer season began to prevail.

1 This letter was published by Barrington in his *Miscellanies*, and was clearly called out by a communication from Barrington himself on the subject with which it deals. ED.

Again; it appears by my Journals for many years past that house-martins retire, to a bird, about the beginning of October; so that a person not very observant of such matters would conclude that they had taken their last farewell; but then it may be seen in my diaries also that considerable flocks have discovered themselves again in the first week of November, and often on the 4th day of that month only for one day; and that not as if they were in actual migration, but playing about at their leisure and feeding calmly, as if no enterprise of moment at all agitated their spirits. And this was the case in the beginning of this very month; for on the 4th of November, more than twenty house-martins, which, in appearance, had all departed about the 7th of October, were seen again for that one morning only sporting between my fields and the Hanger, and feasting on insects which swarmed in that sheltered district. The preceding day was wet and blustering, but the 4th was dark, and mild, and soft, the wind at south-west, and the thermometer at 58′½; a pitch not common at that season of the year. Moreover, it may not be amiss to add in this place, that whenever the thermometer is above 50, the bat comes flitting out in every autumnal and winter month.

Redstart

From all these circumstances, laid together, it is obvious that torpid insects, reptiles, and quadrupeds, are awakened from their profoundest slumbers by a little untimely warmth; and therefore that nothing so much promotes this death-like stupor as a defect of heat. And farther, it is reasonable to suppose that two whole species, or at least many individuals of those two species of British hirundines do never leave this island at all, but partake of the same benumbed state; for we cannot suppose, that after a month's absence, house-martins can return from southern regions to appear for one morning in November, or that house-swallows should leave the districts of Africa to enjoy in March the transient summer of a couple of days.[2]

I am, &c.

2 In such cases the birds on their way south are probably tempted a little north again for a short period by an increased temporary supply of food-stuffs. ED.

At Rogate

Letter 37 also to the Hon. Daines Barrington

Selborne, January 8th, 1778

Dear Sir – There was in this village several years ago a miserable pauper, who from his birth was afflicted with a leprosy, as far as we are aware of a singular kind, since it affected only the palms of his hands and the soles of his feet. This scaly eruption usually broke out twice in the year, at the spring and fall; and, by peeling away, left the skin so thin and tender that neither his hands or feet were able to perform their functions; so that the poor object was half his time on crutches, incapable of employ, and languishing in a tiresome state of indolence and inactivity. His habit was lean, lank, and cadaverous. In this sad plight he dragged on a miserable existence, a burden to himself and his parish which was obliged to support him till he was relieved by death at no more than thirty years of age.

The good women, who love to account for every defect in children by the doctrine of longing, said that his mother felt a violent propensity for oysters, which she was unable to gratify; and that the black rough scurf on his hands and feet were the shells of that fish. We knew his parents, neither of whom were lepers; his father in particular lived to be far advanced in years.

In all ages the leprosy has made dreadful havoc among mankind. The Israelites seem to have been greatly afflicted with it from the most remote times, as appears from the peculiar and repeated injunctions given them in the Levitical law.[1] Nor was the rancour of this foul disorder much abated in the last period of their commonwealth, as may be seen in many passages of the New Testament.

Some centuries ago this horrible distemper prevailed all Europe over; and our forefathers were by no means exempt, as appears by the large provision made for objects labouring under this calamity. There was an hospital for female lepers in the diocese of Lincoln; a noble one near Durham; three in London and Southwark; and perhaps many more in or near our great towns and cities. Moreover, some crowned heads, and other wealthy and charitable personages, bequeathed large legacies to such poor people as languished under this hopeless infirmity.

It must, therefore, in these days be to an humane and thinking person a matter of equal wonder and satisfaction, when he contemplates how nearly this pest is eradicated, and observes that a leper now is a rare sight. He will, moreover, when engaged in such a train of thought naturally inquire for the reason. This happy change, perhaps, may have originated and been continued from the much smaller quantity of salted meat and fish now eaten in these kingdoms; from the use of linen next the skin; from the plenty of better bread; and from the profusion of fruits, roots, legumes, and greens, so common in every family. Three or four centuries ago before there were any enclosures, sown-grasses, field-turnips, or field-carrots, or hay, all the cattle which had grown fat in summer, and were not killed for winter use, were turned out soon after Michaelmas to shift as they could through the dead months; so that no fresh meat could be had in winter or spring. Hence the marvellous account of the vast stores of salted flesh found in the larder of the eldest Spencer[2] in the days of Edward II, even so late in the spring as the 3rd of May. It was from magazines like these that the turbulent barons supported in idleness their riotous swarms of retainers ready for any disorder or mischief. But agriculture is now arrived at such a pitch of perfection that our best and fattest meats are killed in the winter; and no man need eat salted flesh unless he prefers it, that has money to buy fresh.

1 See Leviticus, chap. xiii. and xiv.

2 Viz., Six hundred bacons, eighty carcasses of beef, and six hundred muttons.

One cause of this distemper might be, no doubt, the quantity of wretched fresh and salt fish consumed by the commonalty at all seasons as well as in Lent; which our poor now would hardly be persuaded to touch.

The use of linen changes, shirts or shifts, in the room of sordid and filthy woollen, long worn next the skin, is a matter of neatness comparatively modern; but must prove a great means of preventing cutaneous ails. At this very time woollen, instead of linen, prevails among the poorer Welsh, who are subject to foul eruptions.

The plenty of good wheaten bread that now is found among all ranks of people in the south, instead of that miserable sort which used in old days to be made of barley or beans, may contribute not a little to the sweetening their blood and correcting their juices; for the inhabitants of mountainous districts to this day are still liable to the itch and other cutaneous disorders, from a wretchedness and poverty of diet.

As to the produce of a garden, every middle-aged person of observation may perceive, within his own memory, both in town and country, how vastly the consumption of vegetables is increased. Green-stalls in cities now support multitudes in a comfortable state, while gardeners get fortunes. Every decent labourer also has his garden, which is half his support, as well as his delight; and common farmers provide plenty of beans, peas, and greens, for their hinds to eat with their bacon; and those few that do not are despised for their sordid parsimony, and looked upon as regardless of the welfare of their dependents. Potatoes have prevailed in this little district by means of premiums within these twenty years only; and are much esteemed here now by the poor, who would scarce have ventured to taste them in the last reign.

Our Saxon ancestors certainly had some sort of cabbage, because they call the month of February 'sprout cale'; but long after their days the cultivation of gardens was little attended to. The religious, being men of leisure, and keeping up a constant correspondence with Italy, were the first people among us that had gardens and fruit-trees in any perfection within the wall of their abbeys[3] and priories. The barons neglected every pursuit that did not lead to war or tend to the pleasure of the chase.

3 'In monasteries the lamp of knowledge continued to burn, however dimly. In them men of business were formed for the state: the art of writing was cultivated by the monks; they were the only proficients in mechanics, gardening, and architecture.' – see Dalrymple's *Annals of Scotland*.

It was not till gentlemen took up the study of horticulture themselves that the knowledge of gardening made such hasty advances. Lord Cobham, Lord Ila, and Mr Waller, of Beaconsfield, were some of the first people of rank that promoted the elegant science of ornamenting without despising the superintendence of the kitchen quarters and fruit walls.

A remark made by the excellent Mr Ray, in his *Tour of Europe*, at once surprises us, and corroborates what has been advanced above; for we find him observing, so late as his days, that, 'The Italians use several herbs for sallets, which are not yet, or have not been but lately, used in England, viz., *selleri* (celery), which is nothing else but the sweet smallage; the young shoots whereof, with a little of the head of the root cut off, they eat raw with oil and pepper;' and further adds: 'curled endive blanched is much used beyond seas; and, for a raw sallet, seemed to excel lettuce itself'. Now this journey was undertaken no longer ago than in the year 1663.

I am, &c.

Hartley Mauditt

Letter 38 also to the Hon. Daines Barrington

Selborne, February 12th, 1778

Forte puer, comitum seductus ab agmine fido,
Dixerat, ecquis adest? et, adest, responderat echo,
Hic stupet; utque aciem partes divisit in omnes;
Voce, veni, clamat magna. Vocat ila vocantem.

Dear Sir – In a district so diversified as this, so full of hollow vales and hanging woods, it is no wonder that echoes should abound. Many we have discovered that return the cry of a pack of dogs, the notes of a hunting-horn, a tunable ring of bells, or the melody of birds very agreeably; but we were still at a loss for a polysyllabical articulate echo, till a young gentleman, who had parted from his company in a summer evening walk, and was calling after them, stumbled upon a very curious one in a spot where it might least be expected. At first he was much surprised, and could not be persuaded but that he was mocked by some boy; but repeating his trials in several languages, and finding his respondent to be a very adroit polyglot, he then discerned the deception.

This echo in an evening before rural noises cease, would repeat ten syllables most articulately and distinctly, especially if quick dactyls were chosen. The last syllables of,

Tityre, tu patulae recubans . . .

were as audibly and intelligibly returned as the first; and there is no doubt, could trial have been made, but that at midnight when the air is very elastic, and a dead stillness prevails, one or two syllables more might have been obtained; but the distance rendered so late an experiment very inconvenient.

Quick dactyls, we observed, succeeded best; for when we came to try its powers in slow, heavy, embarrassed spondees of the same number of syllables,

Monstrum horrendum, informe, ingens . . .

we could perceive a return but of four or five.

All echoes have some one place to which they are returned stronger and more distinct than to any other; and that is always the place that lies at right angles with the object of repercussion, and is not too near nor too far off. Buildings, or naked rocks, re-echo much more articulately than hanging woods or vales; because in the latter the voice is, as it were, entangled and embarrassed in the covert, and weakened in the rebound.

The true object of this echo, as we found by various experiments, is the stone-built, tiled hop-kiln in Gally-lane, which measures in front forty feet, and from the ground to the eaves twelve feet. The true *centrum phonicum*, or just distance, is one particular spot in the King's field, in the path to Nore-hill, on the very brink of the steep balk above the hollow cartway. In this case there is no choice of distance; but the path, by mere contingency, happens to be the lucky, the identical spot, because the ground rises or falls so immediately, if the speaker either retires or advances, that his mouth would at once be above or below the object.

We measured this polysyllabical echo with great exactness, and found the distance to fall very short of Dr Plot's rule for distinct articulation; for the Doctor, in his *History of Oxfordshire*, allows a hundred and twenty feet for the return of each syllable distinctly; hence this echo, which gives ten distinct syllables, ought to measure four hundred yards, or one hundred and twenty feet to each syllable; whereas our distance is only two hundred and fifty-eight yards, or near seventy-five feet to each syllable. Thus our measure falls short of the Doctor's, as five to eight; but then it must be acknowledged that this candid philosopher was convinced afterwards, that some latitude must be admitted of in the distance of echoes according to time and place.

When experiments of this sort are making, it should always be remembered that weather and the time of day have a vast influence on an echo; for a dull, heavy, moist air deadens and clogs the sound, and hot sunshine renders the air thin and weak, and deprives it of all its springiness, and a ruffling wind quite defeats the whole. In a still, clear, dewy evening the air is most elastic; and perhaps the later the hour the more so.

Echo has always been so amusing to the imagination, that the poets have personified her; and in their hands she has been the occasion of many a beautiful fiction. Nor need the gravest man be ashamed to appear taken with such a phenomenon, since it may become the subject of philosophical or mathematical inquiries.

One should have imagined that echoes, if not entertaining, must at least have been harmless and inoffensive; yet, Virgil advances a strange notion, that they are injurious to bees. After enumerating some probable and reasonable annoyances, such as prudent owners would wish far removed from their bee-gardens, he adds –

. . . aut ubi concava pulsu
Saxa sonant, vocisque offensa resultat imago.

This wild and fanciful assertion will hardly be admitted by the philosophers of these days, especially as they all now seem agreed that insects are not furnished with any organs of hearing at all![1] But if it should be urged, that though they cannot hear yet perhaps they may feel the repercussions of sounds, I grant it is possible they may. Yet that these impressions are distasteful or hurtful, I deny, because bees, in good summers, thrive well in my outlet, where the echoes are very strong; for this village is another Anathoth, a place of responses and echoes. Besides, it does not appear from experiment that bees are in any way capable of being affected by sounds; for I have often tried my own with a large speaking-trumpet held close to their hives, and with such an exertion of voice as would have hailed a ship at the distance of a mile, and still these insects pursued their various employments undisturbed, and without showing the least sensibility or resentment.

Some time since its discovery this echo is become totally silent, though the object, or hop-kiln, remains; nor is there any mystery in

1 Insects have since been proved to be sensible to sound. Many insects emit musical notes as calls or cries to attract their mates. ED.

The Long Lythe

this defect; for the field between is planted as an hop-garden, and the voice of the speaker is totally absorbed and lost among the poles and entangled foliage of the hops. And when the poles are removed in autumn the disappointment is the same; because a tall, quick-set hedge, nurtured up for the purpose of shelter to the hop ground, entirely interrupts the impulse and repercussion of the voice; so that till these obstructions are removed no more of its garrulity can be expected.

Should any gentleman of fortune think an echo in his park or outlet a pleasing incident, he might build one at little or no expense. For whenever he had occasion for a new barn, stable, dog-kennel, or the like structure, it would be only needful to erect this building on the gentle declivity of an hill, with a like rising opposite to it, at a few hundred yards distance; and perhaps success might be the easier insured could some canal, lake, or stream intervene. From a seat at the *centrum phonicum* he and his friends might amuse themselves sometimes of an evening with the prattle of this loquacious nymph; of whose complacency and decent reserve more may be said than can with truth of every individual of her sex; since she is . . .

. . . quae nec reticere loquenti,
Nec prior ipsa loqui didicit resonabilis echo.

I am, &c.

PS – The classic reader will, I trust, pardon the following lovely quotation, so finely describing echoes, and so poetically accounting for their causes from popular superstition –

Quae bene quom videas, rationem reddere possis
Tute tibi atque aliis, quo pacto per loca sola
Saxa paries formas verborum ex ordine reddant,
Palanteis comites quom monteis inter opacos
Quaerimus, et magna dispersos voce ciemus.
Sex etiam, aut septem loca vidi reddere voces
Unam quom jaceres: ita colles collibus ipsis
Verba repulsantes iterabant dicta referre.
Haec loca capripedes Satyros, Nymphasque tenere
Finitimi fingunt, et Faunos esse loquuntur;
Quorum noctivago strepitu, ludoque jocanti
Adfirmant volga taciturna silentia rumpi,
Chordarumque sonos fieri, dulceisque querelas,
Tibia quas fundit digitis pulsata canentum:
Et genus agricolum late sentiscere, quom Pan
Pinea semiferi capitis velamina quassans,
Unco saepe labro calamos percurrit hianteis,
Fistula silvestrem ne cesset fundere musam.

LUCRETIUS, Lib. iv, l. 576

The Hanger from Gracious Street

Tablet in the garden wall

Letter 39 also to the Hon. Daines Barrington

Selborne, May 13th, 1778

Dear Sir – Among the many singularities attending those amusing birds the swifts, I am now confirmed in the opinion that we have every year the same number of pairs invariably; at least the result of my inquiry has been exactly the same for a long time past. The swallows and martins are so numerous, and so widely distributed over the village, that it is hardly possibly to recount them; while the swifts, though they do not all build in the church, yet so frequently haunt it, and play and rendezvous round it, that they are easily enumerated. The number that I constantly find are eight pairs; about half of which reside in the church, and the rest build in some of the lowest and meanest thatched cottages. Now as these eight pairs, allowance being made for accidents, breed yearly eight pairs more, what becomes annually of this increase; and what determines every spring which pairs shall visit us, and reoccupy their ancient haunts?

Ever since I have attended to the subject of ornithology, I have always supposed that that sudden reverse of affection, that strange ἀντιστοργὴ which immediately succeeds in the feathered kind to the most passionate fondness, is the occasion of an equal dispersion of birds over the face of the earth. Without this provision one favourite district would be crowded with inhabitants, while others would be destitute and forsaken. But the parent birds seem to maintain a jealous superiority, and to oblige the young to seek for new abodes; and the rivalry of the males in many kinds, prevents their crowding the one on the other. Whether the swallows and house-martins return in the same exact number annually is not easy to say, for reasons given above; but it is apparent, as I have remarked before in my Monographies, that the numbers returning bear no manner of proportion to the numbers retiring.[1]

1 Here we get an early hint of that profound problem of multiplication which gave rise later to Malthus's *Theory of Population* and also to the doctrine of the 'struggle for existence', with its Darwinian and Spencerian corollaries of 'natural selection' and the 'survival of the fittest'. It is interesting to observe such first tentative advances, as showing the inevitable trend of thought towards ideas as yet unborn. ED.

Brown Owl

Letter 40 also to the Hon. Daines Barrington

Selborne, June 2nd, 1778

Dear Sir – The standing objection to botany has always been, that it is a pursuit that amuses the fancy and exercises the memory without improving the mind or advancing any real knowledge; and where the science is carried no farther than a mere systematic classification, the charge is but too true. But the botanist that is desirous of wiping off this aspersion should be by no means content with a list of names; he should study plants philosophically, should investigate the laws of vegetation, should examine the powers and virtues of efficacious herbs, should promote their cultivation, and graft the gardener, the planter, and the husbandman, on the phytologist. Not that system is by any means to be thrown aside; without system the field of Nature would be a pathless wilderness: but system should be subservient to, not the main object of, pursuit.[1]

Vegetation is highly worthy of our attention; and in itself is of the utmost consequence to mankind, and productive of many of the

1 See the late voyages to the South Seas.

greatest comforts and elegancies of life. To plants we owe timber, bread, beer, honey, wine, oil, linen, cotton, &c., what not only strengthens our hearts, and exhilarates our spirits, but what secures us from inclemencies of weather and adorns our persons. Man, in his true state of nature, seems to be subsisted by spontaneous vegetation; in middle climes, where grasses prevail, he mixes some animal food with the produce of the field and garden; and it is towards the polar extremes only that, like his kindred bears and wolves, he gorges himself with flesh alone, and is driven, to what hunger has never been known to compel the very beasts, to prey on his own species.[2]

The productions of vegetation have had a vast influence on the commerce of nations, and have been the great promoters of navigation, as may be seen in the articles of sugar, tea, tobacco, opium, ginseng, betel, pepper, &c. As every climate has its peculiar produce, our natural wants bring on a mutual intercourse; so that by means of trade each distant part is supplied with the growth of every latitude. But, without the knowledge of plants and their culture, we must have been content with our hips and haws, without enjoying the delicate fruits of India and the salutiferous drugs of Peru.

Instead of examining the minute distinctions of every various species of each obscure genus, the botanist should endeavour to make himself acquainted with those that are useful. You shall see a man readily ascertain every herb of the field, yet hardly know wheat from barley, or at least one sort of wheat or barley from another.

But of all sorts of vegetation the grasses seem to be most neglected; neither the farmer nor the grazier seem to distinguish the annual from the perennial, the hardy from the tender, nor the succulent and nutritive from the dry and juiceless.

The study of grasses would be of great consequence to a northerly, and grazing kingdom. The botanist that could improve the sward of the district where he lived would be an useful member of society: to raise a thick turf on a naked soil would be worth volumes of systematic knowledge; and he would be the best commonwealth's man that could occasion the growth of 'two blades of grass where one alone was seen before'.

I am, &c.

2 In this pregnant sentence, again, White foreshadows the transition from the age of Linnaeus, bent all on classification, to the age of Darwin, bent all on the interpretation of the facts of nature. ED.

Dorton Cottage

Letter 41 also to the Hon. Daines Barrington

Selborne, July 3rd, 1778

Dear Sir – In a district so diversified with such a variety of hill and dale, aspects, and soils, it is no wonder that great choice of plants should be found. Chalks, clays, sands, sheep-walks and downs, bogs, heaths, woodlands, and champaign fields, cannot but furnish an ample Flora. The deep rocky lanes abound with *filices*, and the pastures and moist woods with *fungi*. If in any branch of botany we may seem to be wanting, it must be in the large aquatic plants, which are not to be expected on a spot far removed from rivers, and lying up amidst the hill country at the spring heads. To enumerate all the plants that have been discovered within our limits would be a needless work; but a short list of the more rare, and the spots where they are to be found, may be neither unacceptable nor unentertaining:

Helleborus foetidus, stinking hellebore, bear's foot, or setterwort – all over the High-wood and Coney-croft-hanger: this continues a great branching plant the winter through, blossoming about January, and is very ornamental in shady walks and shrubberies. The good women give the leaves powdered to children troubled with worms; but it is a violent remedy, and ought to be administered with caution.

Helleborus viridis, green hellebore – in the deep stony lane on the left hand just before the turning to Norton-farm, and at the top of Middle Dorton under the hedge: this plant dies down to the ground early in autumn, and springs again about February, flowering almost as soon as it appears above ground.

Vaccinium oxycoccos, creeping bilberries, or cranberries – in the bogs of Bin's-pond.[1]

Vaccinium myrtillus, whortle, or bleaberries – on the dry hillocks of Wolmer-forest.

Drosera rotundifolia, round-leaved sundew – in the bogs of Bin's-pond.

Drosera longifolia, long-leaved sundew – in the bogs of Bin's-pond.

Comarum palustre, purple comarum, or marsh cinquefoil – in the bogs of Bin's-pond.

Hypericum androsaemum, Tutsan, St John's Wort – in the stony, hollow lanes.

Vinca minor, less periwinkle – in Selborne-hanger and Shrubwood.

Monotropa hypopithys, yellow monotropa, or birds' nest – in Selborne-hanger under the shady beeches, to whose roots it seems to be parasitical, at the north-west end of the Hanger.

Chlora perfoliata, *Blackstonia perfoliata*, *Hudsoni*, perfoliated yellow-wort – on the banks in the King's-field.

Paris quadrifolia, herb of Paris, true-love, or one-berry – in the Church-litten-coppice.

Chrysosplenium oppositifolium, opposite golden saxifrage – in the dark and rocky hollow lanes.

Gentiana amarella, autumnal gentian, or fellwort – on the Zigzag and Hanger.

Lathraea squamaria, tooth-wort – in the Church-litten-coppice under some hazels near the foot-bridge, in Trimming's garden hedge, and on the dry wall opposite Grange-yard.

1 Now drained. ED.

E·H·N·

The Butcher's

Dipsacus pilosus, small teasel – in the Short and Long Lith.

Lathyrus sylvestris, narrow-leaved, or wild lathyrus – in the bushes at the foot of the Short Lith, near the path.

Ophrys spiralis, ladies' traces – in the Long Lith, and towards the south corner of the common.

Ophrys nidus avis, birds' nest ophrys – in the Long Lith under the shady beeches among the dead leaves; in Great Dorton among the bushes, and on the Hanger plentifully.

Serapias latifolia, helleborine – in the High-wood under the shady beeches.

Daphne laureola, spurge laurel – in Selborne-hanger and the Highwood.

Daphne mezereum, the mezereon – in Selborne-hanger among the shrubs, at the south-east end above the cottages.

Lycoperdon tuber, truffles – in the Hanger and High-wood.

Sambucus ebulus, dwarf elder, walwort, or danewort – among the rubbish and ruined foundations of the Priory.[2]

Of all the propensities of plants, none seem more strange than their different periods of blossoming. Some produce their flowers in the winter, or very first dawnings of spring; many when the spring is established; some at midsummer, and some not till autumn. When we see the *helleborus foetidus* and *helleborus niger* blowing at Christmas, the *helleborus hyemalis* in January, and the *helleborus viridis* as soon as ever it emerges out of the ground, we do not wonder, because they are kindred plants that we expect should keep pace the one with the other; but other congenerous vegetables differ so widely in their time of flowering, that we cannot but admire. I shall only instance at present in the *crocus sativus*, the vernal and the autumnal crocus, which have such an affinity, that the best botanists only make them varieties of the same genus, of which there is only one species, not being able to discern any difference in the corolla, or in the internal structure. Yet the vernal crocus expands its flowers

2 In the first edition this letter ended here; but in the quarto edited by Mitford the following passage was added to it. The additional paragraph has appeared in all the subsequent editions which I have consulted. I do not know whence Mitford derived it, nor on what authority he added it in this particular position. ED.

by the beginning of March at farthest, and often in very rigorous weather; and cannot be retarded but by some violence offered; while the autumnal (the saffron) defies the influence of the spring and summer, and will not blow till most plants begin to fade and run to seed. This circumstance is one of the wonders of the creation, little noticed because a common occurrence; yet ought not to be overlooked on account of its being familiar, since it would be as difficult to be explained as the most stupendous phenomenon in nature.

Say what impels, amidst surrounding snow
Congeal'd, the crocus' flamy bud to glow?
Say, what retards, amidst the summer's blaze
Th' autumnal bulb, till pale, declining days?
The God of Seasons; whose pervading power
Controls the sun, or sheds the fleecy shower:
He bids each flower his quickening word obey,
Or to each lingering bloom enjoins delay.

Woodcock

Letter 42 also to the Hon. Daines Barrington

Omnibus animalibus reliquis certus et uniusmodi, et in suo cuique genere
incessus est: aves solae vario meatu feruntur, et in terra, et in aere.

Selborne, August 7th, 1778

Dear Sir – A good ornithologist should be able to distinguish birds by their air as well as by their colours and shape; on the ground as well as on the wing; and in the bush as well as in the hand. For, though it must not be said that every species of birds has a manner peculiar to itself, yet there is somewhat in most *genera* at least, that at first sight discriminates them, and enables a judicious observer to pronounce upon them with some certainty. Put a bird in motion

Et vera incessu patuit . . . [1]

Thus kites and buzzards sail round in circles with wings expanded

1 The original edition reads *vera*, with a needless circumflex clearly due to a printer's blunder. In this and many other cases I have not thought it necessary slavishly to reproduce the particular vagaries of Mr Benjamin White's compositors. Either Gilbert White did not correct his own proofs, or, if he corrected them, allowed many foolish errors of the printer to pass unnoticed. ED

and motionless; and it is from their gliding manner that the former are still called in the north of England gleads, from the Saxon verb glidan, to glide. The kestrel, or windover, has a peculiar mode of hanging in the air in one place, his wings all the while being briskly agitated. Hen-harriers fly low over heaths or fields of corn, and beat the ground regularly like a pointer or setting-dog. Owls move in a buoyant manner, as if lighter than the air; they seem to want ballast. There is a peculiarity belonging to ravens that must draw the attention even of the most incurious – they spend all their leisure time in striking and cuffing each other on the wing in a kind of playful skirmish; and, when they move from one place to another, frequently turn on their backs with a loud croak, and seem to be falling to the ground. When this odd gesture betides them, they are scratching themselves with one foot, and thus lose the centre of gravity. Rooks sometimes dive and tumble in a frolicksome manner; crows and daws swagger in their walk; woodpeckers fly *volatu undoso*, opening and closing their wings at every stroke, and so are always rising or falling in curves. All of this genus use their tails, which incline downward, as a support while they run up trees. Parrots, like all other hooked-clawed birds, walk awkwardly, and make use of their bill as a third foot, climbing and descending with ridiculous caution. All the *gallinae* parade and walk gracefully, and run nimbly; but fly with difficulty, with an impetuous whirring, and in a straight line. Magpies and jays flutter with powerless wings, and make no dispatch; herons seem incumbered with too much sail for their light bodies, but these vast hollow wings are necessary in carrying burdens, such as large fishes and the like; pigeons, and particularly the sort called smiters, have a way of clashing their wings the one against the other over their backs with a loud snap; another variety, called tumblers, turn themselves over in the air. Some birds have movements peculiar to the season of love: thus ringdoves, though strong and rapid at other times, yet in the spring hang about on the wing in a toying and playful manner; thus the cock-snipe, while breeding, forgetting his former flight, fans the air like the windhover; and the green-finch in particular, exhibits such languishing and faltering gestures as to appear like a wounded and dying bird; the kingfisher darts along like an arrow; fern-owls, or goat-suckers, glance in the dusk over the tops of trees like a meteor; starlings as it were swim along, while missel-thrushes use a wild and desultory flight; swallows sweep over the surface of the ground and water, and distinguish themselves by rapid turns and quick evolutions; swifts dash round in circles; and the bank-martin moves with

The Wakes in Bell's time

frequent vacillations like a butterfly. Most of the small birds fly by jerks, rising and falling as they advance. Most small birds hop; but wagtails and larks walk, moving their legs alternately. Skylarks rise and fall perpendicularly as they sing; woodlarks hang poised in the air; and titlarks rise and fall in large curves, singing in their descent. The white-throat uses odd jerks and gesticulations over the tops of hedges and bushes. All the duck-kind waddle; divers and auks walk as if fettered, and stand erect on their tails: these are the *compedes* of Linnaeus. Geese and cranes, and most wild fowls, move in figured flights, often changing their position. The secondary *remiges* of Tringae, wild ducks, and some others, are very long, and give their wings, when in motion, an hooked appearance. Dabchicks, moorhens, and coots fly erect, with their legs hanging down, and hardly make any dispatch; the reason is plain, their wings are placed too forward out of the true centre of gravity; as the legs of auks and divers are situated too backward.

Sparrow-hawk

Letter 43 also to the Hon. Daines Barrington[1]

Selborne, September 9th, 1778

Dear Sir – From the motion of birds, the transition is natural enough to their notes and language, of which I shall say something. Not that I would pretend to understand their language like the vizier; who, by the recital of a conversation which passed between two owls reclaimed a sultan,[2] before delighting in conquest and devastation; but I would be thought only to mean that many of the winged tribes have various sounds and voices adapted to express their various passions, wants, and feelings; such as anger, fear, love, hatred, hunger, and the like. All species are not equally eloquent; some are copious and fluent as it were in their utterance, while others are confined to a few important sounds; no bird, like the fish kind,[3] is quite mute, though some are rather silent. The language of

1 See *Spectator*, Vol. vii, No. 512.
2 This is clearly not a real letter, but an additional essay on the notes of birds, written when the idea of publication had been adopted. ED.
3 A few fish utter cries. The grey gurnard grunts loud enough to be heard at a considerable distance. As a rule, however, fish are 'somewhat silent'. ED.

birds is very ancient, and, like other ancient modes of speech, very elliptical;[4] little is said, but much is meant and understood.

The notes of the eagle-kind are shrill and piercing; and about the season of nidification much diversified, as I have been often assured by a curious observer of Nature, who long resided at Gibraltar, where eagles abound. The notes of our hawks much resemble those of the king of birds. Owls have very expressive notes; they hoot in a fine vocal sound, much resembling the *vox humana*, and reducible by a pitch-pipe to a musical key. This note seems to express complacency and rivalry among the males; they use also a quick call and an horrible scream; and can snore and hiss when they mean to menace. Ravens, besides their loud croak, can exert a deep and solemn note that makes the woods to echo; the amorous sound of a crow is strange and ridiculous; rooks, in the breeding season, attempt sometimes in the gaiety of their hearts to sing, but with no great success; the parrot-kind have many modulations of voice, as appears by their aptitude to learn human sounds; doves coo in an amorous and mournful manner, and are emblems of despairing lovers; the woodpecker sets up a sort of loud and hearty laugh; the fern-owl, or goat-sucker, from the dusk till day-break, serenades his mate with the clattering of castanets.[5] All the tuneful *passeres* express their complacency by sweet modulations, and a variety of melody. The swallow, as has been observed in a former letter, by a shrill alarm bespeaks the attention of the other hirundines, and bids them be aware the hawk is at hand. Aquatic and gregarious birds; especially the nocturnal, that shift their quarters in the dark, are very noisy and loquacious; as cranes, wild-geese, wild-ducks, and the like; their perpetual clamour prevents them from dispersing and losing their companions.

In so extensive a subject, sketches and outlines are as much as can be expected; for it would be endless to instance in all the infinite variety of the feathered nation. We shall therefore confine the remainder of this letter to the few domestic fowls of our yards, which are most known, and therefore best understood. And first the peacock, with his gorgeous train, demands our attention; but, like most of the gaudy birds, his notes are grating and shocking to the

4 This is a true and deep remark – one of White's many anticipatory *aperçus*. Later research has shown that very early human speech, and the speech of very undeveloped races, is elliptical in the extreme. ED.

5 The fern-owl, or nightjar, utters a note which White here sadly underestimates. Though not musical, it is full of profound and weird emotion. ED.

ear: the yelling of cats, and the braying of an ass, are not more disgustful. The voice of the goose is trumpet-like, and clanking; and once saved the Capitol at Rome, as grave historians assert: the hiss, also, of the gander, is formidable and full of menace, and 'protective of his young'. Among ducks the sexual distinction of voice is remarkable; for, while the quack of the female is loud and sonorous, the voice of the drake is inward and harsh, and feeble, and scarce discernible. The cock turkey struts and gobbles to his mistress in a most uncouth manner; he hath also a pert and petulant note when he attacks his adversary. When a hen turkey leads forth her young brood she keeps a watchful eye; and if a bird of prey appear, though ever so high in the air, the careful mother announces the enemy with a little inward moan, and watches him with a steady and attentive look; but, if he approach, her note becomes earnest and alarming, and her outcries are redoubled.

No inhabitants of a yard seem possessed of such a variety of expression and so copious a language as common poultry. Take a chicken of four or five days old, and hold it up to a window where there are flies, and it will immediately seize its prey, with little twitterings of complacency; but if you tender it a wasp or a bee, at once its note becomes harsh, and expressive of disapprobation and a sense of danger. When a pullet is ready to lay she intimates the event by a joyous and easy soft note. Of all the occurrences of their life that of laying seems to be the most important; for no sooner has a hen disburdened herself, than she rushes forth with a clamorous kind of joy, which the cock and the rest of his mistresses immediately adopt. The tumult is not confined to the family concerned, but catches from yard to yard, and spreads to every homestead within hearing, till at last the whole village is in an uproar. As soon as a hen becomes a mother her new relation demands a new language: she then runs clucking [clacking] and screaming about, and seems agitated as if possessed. The father of the flock has also a considerable vocabulary; if he finds food, he calls a favourite concubine to partake; and if a bird of prey passes over, with a warning voice he bids his family beware. The gallant chanticleer has, at command, his amorous phrases and his terms of defiance. But the sound by which he is best known is his crowing: by this he has been distinguished in all ages as the countryman's clock or larum, as the watchman that proclaims the divisions of the night. Thus the poet elegantly styles him:

> . . . the crested cock, whose clarion sounds
> The silent hours.

The Wakes in Bell's time

A neighbouring gentleman one summer had lost most of his chickens by a sparrow-hawk, that came gliding down between a faggot pile and the end of his house to the place where the coops stood. The owner, inwardly vexed to see his flock thus diminished, hung a setting-net adroitly between the pile and the house, into which the caitiff dashed, and was entangled. Resentment suggested the law of retaliation; he therefore clipped the hawk's wings, cut off his talons, and, fixing a cork on his bill, threw him down among the brood-hens. Imagination cannot paint the scene that ensued; the expressions that fear, rage, and revenge, inspired, were new, or at least such as had been unnoticed before: the exasperated matrons upbraided, they execrated, they insulted, they triumphed. In a word, they never desisted from buffeting their adversary till they had torn him in an hundred pieces.

Letter 44 also to the Hon. Daines Barrington

. . . Monstrent

Quid tantum Oceano properent se tingere soles
Hyberni; vel quae tardis mora noctibus obstet.

Selborne[1]

Gentlemen who have outlets might contrive to make ornament subservient to utility: a pleasing eye-trap might also contribute to promote science: an obelisk in a garden or park might be both an embellishment and an heliotrope.

Any person that is curious, and enjoys the advantage of a good horizon, might, with little trouble, make two heliotropes; the one for the winter, the other for the summer solstice: and the two erections might be constructed with very little expense; for two pieces of timber frame-work, about ten or twelve feet high, and four feet broad at the base, and close lined with plank, would answer the purpose.

1 This letter has no date of time, but one of place only. It, and most of those which follow it, were not, I believe, ever really written to Barrington. They are notes called forth by the subjects of the previous series. Many of them are not dated at all: these, I fancy, were written merely to embody other important observations not alluded to in the genuine correspondence. Their style is accordingly more 'literary' and less spontaneous: they are therefore of far inferior interest and importance to the actual letters. ED.

The erection for the former should, if possible, be placed within sight of some window in the common sitting-parlour; because men, at that dead season of the year, are usually within doors at the close of the day; while that for the latter might be fixed for any given spot in the garden or outlet: whence the owner might contemplate, in a fine summer's evening, the utmost extent that the sun makes to the northward at the season of the longest days. Now nothing would be necessary but to place these two objects with so much exactness, that the westerly limb of the sun, at setting, might but just clear the winter heliotrope to the west of it on the shortest day, and that the whole disc of the sun, at the longest day, might exactly at setting also clear the summer heliotrope to the north of it.

By this simple expedient it would soon appear that there is no such thing, strictly speaking, as a solstice; for, from the shortest day, the owner would, every clear evening, see the disc advancing, at its setting, to the westward of the object; and, from the longest day observe the sun retiring backwards every evening at its setting, towards the object westward, till, in a few nights, it would set quite behind it and so by degrees, to the west of it: for when the sun comes near the summer solstice, the whole disc of it would at first set behind the object; after a time the northern limb would first appear, and so every night gradually more, till at length the whole diameter would set northward of it for about three nights; but on the middle night of the three, sensibly more remote than the former or following. When beginning its recess from the summer tropic, it would continue more and more to be hidden every night, till at length it would descend quite behind the object again; and so nightly more and more to the westward.

Letter 45 also to the Hon. Daines Barrington

. . . .Mugire vedebis
Sub pedibus terram, et descendere montibus ornos.

Selborne

When I was a boy I used to read, with astonishment and implicit assent, accounts in *Baker's Chronicle* of walking hills and travelling mountains. John Philips, in his *Cyder*, alludes to the credit that was given to such stories with a delicate but quaint vein of humour peculiar to the author of the *Splendid Shilling.*

> I nor advise, nor reprehend the choice
> Of Marcley Hill; the apple nowhere finds
> A kinder mould; yet 'tis unsafe to trust
> Deceitful ground: who knows but that once more
> This mount may journey, and his present site
> Forsaken, to thy neighbour's bounds transfer
> Thy goodly plans, affording matter strange
> For law debates.

But, when I came to consider better, I began to suspect that though our hills may never have journeyed far, yet that the ends of many of them have slipped and fallen away at distant periods, leaving the cliffs bare and abrupt. This seems to have been the case with Nore and Whetham Hills; and especially with the ridge between Harteley Park and Ward-le-Ham, where the ground has slid into vast swellings and furrows; and lies still in such romantic confusion as cannot be accounted for from any other cause. A strange event, that happened not long since, justifies our suspicions; which, though it befell not within the limits of this parish, yet as it was within the hundred of Selborne, and as the circumstances were singular, may fairly claim a place in a work of this nature.

The months of January and February, in the year 1774, were remarkable for great melting snows and vast gluts of rain; so that by the end of the latter month the land-springs, or lavants, began to prevail, and to be near as high as in the memorable winter of 1764 The beginning of March also went on in the same tenor; when, in the night between the eighth and ninth of that month, a considerable part of the great woody hanger at Hawkley was torn from its place, and fell down, leaving a high free-stone cliff naked and bare, and resembling the steep side of a chalk-pit. It appears that this huge fragment, being perhaps sapped and undermined by waters, foundered, and was ingulfed, going down in a perpendicular direction; for a gate which stood in the field, on the top of the hill, after sinking with its posts for thirty or forty feet, remained in so true and upright a position as to open and shut with great exactness, just as in its first situation. Several oaks also are still standing, and in a state of vegetation after taking the same desperate leap. That great part of this prodigious mass was absorbed in some gulf below, is plain also from the inclining ground at the bottom of the hill, which is free and unincumbered; but would have been buried in heaps of rubbish, had the fragment parted and fallen forward. About an hundred yards from the foot of this hanging coppice stood a cottage by the side of a lane; and two hundred yards lower, on the other side of the lane, was a farmhouse, in which lived a labourer and his family; and, just by, a stout new barn The cottage was inhabited by an old woman and her son, and his wife. These people in the evening, which was very dark and tempestuous, observed that the brick floors of their kitchen began to heave and part; and that the walls seemed to open, and the roofs to crack: but they all agree that no tremor of the ground, indicating an earthquake, was ever felt; only that the wind continued to make a most tremendous roaring in the woods and hangers. The miserable inhabitants, not daring to go to bed, remained in the utmost solicitude and confusion, expecting every moment to be buried under the ruins of their shattered edifices. When daylight came they were at leisure to contemplate the devastations of the night: they then found that a deep rift, or chasm, had opened under their houses, and torn them, as it were, in two; and that one end of the barn had suffered in a similar manner; that a pond near the cottage had undergone a strange reverse, becoming deep at the shallow end, and so *vice versa*; that many large oaks were removed out of their perpendicular, some thrown down, and some fallen into the heads of neighbouring trees; and that a gate was thrust forward, with its hedge, full six feet, so as to require a new track to be made to it. From the foot of the cliff the general course of the ground, which is pasture, inclines in a moderate descent for half a mile, and is interspersed with some hillocks, which were rifted, in every direction, as well towards the great woody hanger, as from it. In the first pasture the deep clefts began; and running across the lane, and under the buildings, made such vast shelves that the road was impassable for some time; and so over to an arable field on the other side, which was strangely torn and disordered. The second pasture field, being more soft and springy, was protruded forward without many fissures in the turf, which was raised in long ridges resembling graves, lying at right angles to the motion. At the bottom of this enclosure the soil and turf rose many feet against the bodies of some oaks that obstructed their farther course, and terminated this awful commotion.

The perpendicular height of the precipice in general is twenty three yards; the length of the lapse or slip as seen from the fields below, one hundred and eighty-one; and a partial fall, concealed in the coppice, extends seventy yards more; so that the total length of this fragment that fell was two hundred and fifty-one yards. About fifty acres of land suffered from this violent convulsion; two houses were entirely destroyed; one end of a new barn was left in ruins, the walls being cracked through the very stones that composed them; a hanging coppice was changed to a naked rock; and some grass grounds and an arable field so broken and rifted by the chasms as to be rendered for a time neither fit for the plough or safe for pasturage, till considerable labour and expense had been bestowed in levelling the surface and filling in the gaping fissures.[1]

1 In this letter White shows himself prophetic of Lyell's famous doctrine that geological phenomena are due, not to mighty cataclysms, but to the slow result of causes still in action. ED.

Field-crickets

Letter 46 also to the Hon. Daines Barrington

resonant arbusta . . .

Selborne

There is a steep abrupt pasture field interspersed with furze close to the back of this village, well known by the name of Short Lithe, consisting of a rocky dry soil, and inclining to the afternoon sun. This spot abounds with the *gryllus campestris*, or field-cricket; which, though frequent in these parts, is by no means a common insect in many other countries.

As their cheerful summer cry cannot but draw the attention of a naturalist, I have often gone down to examine the economy of these *grylli*, and study their mode of life; but they are so shy and cautious that it is no easy matter to get a sight of them; for feeling a person's footsteps as he advances, they stop short in the midst of their song, and retire backward nimbly into their burrows, where they lurk till all suspicion of danger is over.

At first we attempted to dig them out with a spade, but without any great success; for either we could not get to the bottom of the hole, which often terminated under a great stone; or else in breaking up the ground we inadvertently squeezed the poor insect to death. Out of one so bruised we took a multitude of eggs, which were long and narrow, of a yellow colour, and covered with a very tough skin. By this accident we learned to distinguish the male from the female; the former of which is shining black, with a golden stripe across his shoulders; the latter is more dusky, more capacious about the abdomen, and carries a long, sword-shaped weapon at her tail, which probably is the instrument with which she deposits her eggs in crannies and safe receptacles.

Where violent methods will not avail, more gentle means will often succeed, and so it proved in the present case; for, though a spade be too boisterous and rough an implement, a pliant stalk of grass, gently insinuated into the caverns, will probe their windings to the bottom, and quickly bring out the inhabitant; and thus the humane inquirer may gratify his curiosity without injuring the object of it. It is remarkable, that though these insects are furnished with long legs behind, and brawny thighs for leaping, like grasshoppers; yet when driven from their holes they show no activity, but crawl along in a shiftless manner, so as easily to be taken; and again, though provided with a curious apparatus of wings, yet they never exert them when there seems to be the greatest occasion. The males only make that shrilling noise, perhaps, out of rivalry and emulation, as is the case with many animals which exert some sprightly note during their breeding time. It is raised by a brisk friction of one wing against the other. They are solitary beings, living singly male and female, each as it may happen; but there must be a time when the sexes have some intercourse, and then the wings may be useful perhaps during the hours of night. When the males meet they will fight fiercely, as I found by some which I put into the crevices of a dry stone wall, where I should have been glad to have made them settle. For though they seemed distressed by being taken out of their knowledge, yet the first that got possession of the chinks would seize on any that were intruded upon them with a vast row of serrated fangs. With their strong jaws, toothed like the shears of a lobster's claws, they perforate and round their curious regular cells, having no fore-claws to dig, like the mole-cricket. When taken in hand I could not but wonder that they never offered to defend themselves, though armed with such formidable weapons. Of such herbs as grow before the mouths of their burrows they eat indiscriminately, and on a little platform which they make just by, they drop their dung; and never, in the day time, seem to stir more than two or three inches from home. Sitting in the entrance of their caverns they chirp all night as well as day from the middle of the month of May to the middle of July; and in hot weather, when they are most vigorous, they make the hills echo, and in the stiller hours of darkness may be heard to a considerable distance. In the beginning of the season their notes are more faint and inward; but become louder as the summer advances, and so die away again by degrees.

Sounds do not always give us pleasure according to their sweetness and melody; nor do harsh sounds always displease. We are more apt to be captivated or disgusted with the associations which they promote than with the notes themselves. Thus the shrilling of the field-cricket, though sharp and stridulous, yet marvellously delights some hearers, filling their minds with a train of summer ideas of everything that is rural, verdurous, and joyous.

About the 10th of March the crickets appear at the mouths of their cells, which they then open and bore, and shape very elegantly. All that ever I have seen at that season were in their pupa state, and had only the rudiments of wings, lying under a skin or coat, which must be cast before the insect can arrive at its perfect state[1] from whence I should suppose that the old ones of last year do not always survive the winter. In August their holes begin to be obliterated, and the insects are seen no more till spring.

Not many summers ago I endeavoured to transplant a colony to the terrace in my garden, by boring deep holes in the sloping turf. The new inhabitants stayed some time, and fed and sung; but wandered away by degrees, and were heard at a farther distance every morning, so that it appears that on this emergency they made use of their wings in attempting to return to the spot from which they were taken.

One of these crickets when confined in a paper cage and set in the sun, and supplied with plants moistened with water, will feed and thrive, and become so merry and loud as to be irksome in the same room where a person is sitting; if the plants are not wetted it will die.

1 We have observed that they cast these skins in April, which are then seen lying at the mouths of their holes.

House-cricket

Letter 47 also to the Hon. Daines Barrington

Selborne

Dear Sir –

... Far from all resort of mirth
Save the cricket on the hearth.
MILTON'S *Il Penseroso*

While many other insects must be sought after in fields and woods, and waters, the *gryllus domesticus*, or house-cricket, resides altogether within our dwellings, intruding itself upon our notice whether we will or no. This species delights in new-built houses, being, like the spider, pleased with the moisture of the walls; and besides, the softness of the mortar enables them to burrow and mine between the joints of the bricks or stones, and to open communications from one room to another. They are particularly fond of kitchens and bakers' ovens, on account of their perpetual warmth.

Tender insects that live abroad either enjoy only the short period of one summer, or else doze away the cold uncomfortable months in profound slumbers; but these, residing as it were in a torrid zone, are always alert and merry – a good Christmas fire is to them like the heats of the dog-days. Though they are frequently heard by day, yet is their natural time of motion only in the night. As soon as it grows dusk, the chirping increases, and they come running forth, and are from the size of a flea to that of their full stature. As one should suppose, from the burning atmosphere which they inhabit they are a thirsty race, and show a great propensity for liquids, being found frequently drowned in pans of water, milk, broth, or the like. Whatever is moist they affect; and therefore often gnaw holes in wet woollen stockings and aprons that are hung to the fire;

they are the housewife's barometer, foretelling her when it will rain, and are prognostic sometimes, she thinks, of ill or good luck, of the death of a near relation, or the approach of an absent lover. By being the constant companions of her solitary hours they naturally become the objects of her superstition. These crickets are not only very thirsty, but very voracious; for they will eat the scummings of pots, and yeast, salt, and crumbs of bread, and any kitchen offal or sweepings. In the summer we have observed them to fly when it became dusk out of the windows and over the neighbouring roofs. This feat of activity accounts for the sudden manner in which they often leave their haunts, as it does for the method by which they come to houses where they were not known before. It is remarkable that many sorts of insects seem never to use their wings, but when they have a mind to shift their quarters and settle new colonies. When in the air they move '*volatu undoso*', in waves or curves, like woodpeckers, opening and shutting their wings at every stroke, and so are always rising or sinking.

When they increase to a great degree, as they did once in the house where I am now writing, they become noisome pests, flying into the candles, and dashing into people's faces; but may be blasted and destroyed by gunpowder discharged into their crevices and crannies. In families at such times they are like Pharaoh's plague of frogs – 'in their bedchambers, and upon their beds, and in their ovens, and in their kneading troughs'.[1] Their shrilling noise is occasioned by a brisk attrition of their wings. Cats catch hearth crickets, and, playing with them as they do with mice, devour them. Crickets may be destroyed, like wasps, by phials filled with beer, or any liquid, and set in their haunts; for being always eager to drink, they will crowd in till the bottles are full.

1 Exodus viii: 3.

Mole-cricket

Letter 47 also to the Hon. Daines Barrington

Selborne

How diversified are the modes of life not only of incongruous but even of congenerous animals; and yet their specific distinctions are not more various than their propensities. Thus while the field-cricket delights in sunny dry banks, and the house-cricket rejoices amidst the glowing heat of the kitchen hearth or oven, the *Gryllus gryllotalpa* (the mole-cricket), haunts moist meadows, and frequents the sides of ponds and banks of streams, performing all its functions in a swampy wet soil. With a pair of fore-feet curiously adapted to the purpose, it burrows and works under ground like the mole, raising a ridge as it proceeds, but seldom throwing up hillocks.

As mole-crickets often infest gardens by the side of canals, they are unwelcome guests to the gardener, raising up ridges in their subterraneous progress, and rendering the walks unsightly. If they take to the kitchen quarters they occasion great damage among the plants and roots, by destroying whole beds of cabbages, young legumes, and flowers. When dug out they seem very slow and helpless, and make no use of their wings by day; but at night they come abroad, and make long excursions, as I have been convinced by finding stragglers, in a morning, in improbable places. In fine weather, about the middle of April, and just at the close of day, they begin to solace themselves with a low, dull, jarring note, continued for a long time without interruption, and not unlike the chattering of the fern-owl, or goat-sucker, but more inward.[1]

About the beginning of May they lay their eggs, as I was once an eye-witness; for a gardener at an house where I was on a visit, happening to be mowing, on the 6th of that month, by the side of a canal, his scythe struck too deep, pared off a large piece of turf, and laid open to view a curious scene of domestic economy :

> *... Ingentem lato dedit ore fenestram:*
> *Apparet domus intus, et atria longa patescunt:*
> *Apparent ... penetralia.*

There were many caverns and winding passages leading to a kind of chamber, neatly smoothed and rounded, and about the size of a moderate snuff-box. Within this secret nursery were deposited near an hundred eggs of a dirty yellow colour, and enveloped in a tough skin, but too lately excluded to contain any rudiments of young, being full of a viscous substance. The eggs lay but shallow, and within the influence of the sun, just under a little heap of fresh moved mould, like that which is raised by ants.

When mole-crickets fly they move '*cursu undoso*', rising and falling in curves, like the other species mentioned before. In different parts of this kingdom people call them fen-crickets, churr-worms, and eve-churrs,[2] all very apposite names.

Anatomists, who have examined the intestines of these insects astonish me with their accounts; for they say that, from the structure, position, and number of their stomachs, or maws, there seems to be good reason to suppose that this and the two former species ruminate or chew the cud like many quadrupeds!

1 Its note still more strikingly resembles that of the grasshopper-warbler. ED.

2 White, I think, is mistaken in supposing that the word *eve-churr* refers to the mole-cricket. It is a variant on the name nightjar, now commonly applied to the fern-owl, or goat-sucker. In the form of *eve-jar* it has been introduced into literature by Mr George Meredith in his exquisite poem, 'Love in a Valley'. ED.

Black-winged Stilt

Letter 49 also to the Hon. Daines Barrington

Selborne, May 7th, 1779

It is now more than forty years that I have paid some attention to the ornithology of this district, without being able to exhaust the subject: new occurrences still arise as long as any inquiries are kept alive.

In the last week of last month five of those most rare birds, too uncommon to have obtained an English name, but known to naturalists by the terms of *himantopus*, or *loripes*, and *charadrius himantopus*,[1] were shot upon the verge of Frinsham-pond [Frensham-pond], a large lake belonging to the Bishop of Winchester, and lying between Wolmer-forest and the town of Farnham, in the county of Surrey. The pond keeper says there were three brace in the flock: but, that after he had satisfied his curiosity, he suffered the sixth to remain unmolested. One of these specimens I procured, and found the length of the legs to be so extraordinary, that, at first

1 This bird is apparently the *Himantopus melanopterus* of modern ornithology. ED.

sight, one might have supposed the shanks had been fastened on to impose on the credulity of the beholder: they were legs in *caricatura*; and had we seen such proportions on a Chinese or Japan screen we should have made large allowances for the fancy of the draughtsman. These birds are of the plover family, and might with propriety be called the stilt-plovers. Brisson, under that idea, gives them the apposite name of *l'échasse*. My specimen, when drawn and stuffed with pepper, weighed only four ounces and a quarter, though the naked part of the thigh measured three inches and a half, and the legs four inches and a half. Hence we may safely assert that these birds exhibit, weight for inches, incomparably the greatest length of legs of any known bird. The flamingo, for instance, is one of the most long-legged birds, and yet it bears no manner of proportion to the *himantopus*; for a cock flamingo weighs, at an average, about four pounds avoirdupois; and his legs and thighs measure usually about twenty inches. But four pounds are fifteen times and a fraction more than four ounces, and one quarter; and if four ounces and a quarter have eight inches of legs, four pounds must have one hundred and twenty inches and a fraction of legs; viz., somewhat more than ten feet; such a monstrous proportion as the world never saw! If you should try the experiment in still larger birds the disparity would still increase. It must be matter of great curiosity to see the stilt-plover move; to observe how it can wield such a length of lever with such feeble muscles as the thighs seem to be furnished with. At best one should expect it to be but a bad walker; but what adds to the wonder is, that it has no back toe. Now without that steady prop to support its steps, it must be liable, in speculation, to perpetual vacillations, and seldom able to preserve the true centre of gravity.

The old name of *himantopus* is taken from Pliny; and, by an awkward metaphor, implies that the legs are as slender and pliant as if cut out of a thong of leather. Neither Willughby nor Ray, in all their curious researches, either at home or abroad, ever saw this bird. Mr Pennant never met with it in all Great Britain, but observed it often in the cabinets of the curious at Paris. Hasselquist says that it migrates to Egypt in the autumn; and a most accurate observer of Nature[2] has assured me that he has found it on the banks of the streams in Andalusia.

2 The 'accurate observer of nature' so often alluded to is almost undoubtedly Gilbert White's brother, the Revd John White of Gibraltar. ED.

The church & yew hedge

Our writers record it to have been found only twice in Great Britain. From all these relations it plainly appears that these long-legged plovers are birds of South Europe, and rarely visit our island; and when they do, are wanderers and stragglers, and impelled to make so distant and northern an excursion from motives or accidents for which we are not able to account. One thing may fairly be deduced, that these birds come over to us from the Continent, since nobody can suppose that a species not noticed once in an age, and of such a remarkable make, can constantly breed unobserved in this kingdom.

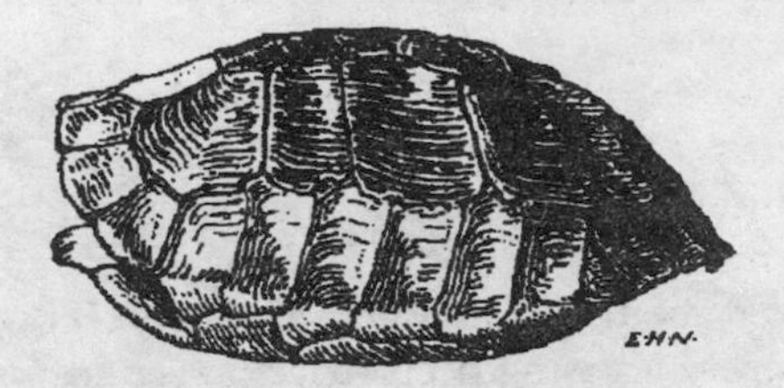

White's Tortoise's Shell

Letter 50 also to the Hon. Daines Barrington

Selborne, April 21st, 1780

Dear Sir – The old Sussex tortoise, that I have mentioned to you so often, is become my property. I dug it out of its winter dormitory in March last, when it was enough awakened to express its resentments by hissing; and, packing it in a box with earth, carried it eighty miles in post-chaises. The rattle and hurry of the journey so perfectly roused it that, when I turned it out on a border, it walked twice down to the bottom of my garden; however, in the evening, the weather being cold, it buried itself in the loose mould, and continues still concealed.

As it will be under my eye, I shall now have an opportunity of enlarging my observations on its mode of life, and propensities; and perceive already that, towards the time of coming forth, it opens a breathing place in the ground near its head, requiring, I conclude, a freer respiration as it becomes more alive. This creature not only goes under the earth from the middle of November to the middle of April, but sleeps great part of the summer; for it goes to bed in the longest days at four in the afternoon, and often does not stir in the morning till late. Besides, it retires to rest for every shower, and does not move at all in wet days.

When one reflects on the state of this strange being, it is a matter of wonder to find that Providence should bestow such a profusion of days, such a seeming waste of longevity, on a reptile that appears to relish it so little as to squander more than two-thirds of its existence in a joyless stupor, and be lost to all sensation for months together in the profoundest of slumbers

While I was writing this letter, a moist and warm afternoon, with the thermometer at 50, brought forth troops of shell-snails; and at the same juncture, the tortoise heaved up the mould and put out its head; and the next morning came forth, as it were raised from the dead, and walked about till four in the afternoon. This was a curious coincidence! a very amusing occurrence! to see such a similarity of feelings between the two φερέοικοι! for so the Greeks called both the shell-snail and the tortoise.

Summer birds are, this cold and backward spring, unusually late: I have seen but one swallow yet. This conformity with the weather convinces me more and more that they sleep in the winter![1]

1 Still harking back to the same old error. ED.

The Queen's Arms

Letter 51 also to the Hon. Daines Barrington

Selborne, September 3rd, 1781

I have now read your *Miscellanies* through with much care and satisfaction; and am to return you my best thanks for the honourable mention made in them of me as a naturalist, which I wish I may deserve.

In some former letters I expressed my suspicions that many of the house-martins do not depart in the winter far from this village. I therefore determined to make some search about the south-east end of the hill, where I imagined they might slumber out the uncomfortable months of winter. But supposing that the examination would be made to the best advantage in the spring, and observing that no martins had appeared by the 11th of April last; on that day I employed some men to explore the shrubs and cavities of the suspected spot. The persons took pains, but without any success; however, a remarkable incident occurred in the midst of our pursuit; while the labourers were at work, a house-martin, the first that had been seen this year, came down the village in the sight of several people, and went at once into a nest, where it stayed a short time, and then flew over the houses; for some days after no martins were observed, not till the 16th of April, and then only a pair. Martins in general were remarkably late this year.

The Hermitage

Letter 52 also to the Hon. Daines Barrington

Selborne, September 9th, 1781

I have just met with a circumstance respecting swifts, which furnishes an exception to the whole tenor of my observations ever since I have bestowed any attention on that species of hirundines. Our swifts, in general, withdrew this year about the first day of August, all save one pair, which in two or three days was reduced to a single bird. The perseverance of this individual made me suspect that the strongest of motives, that of an attachment to her young, could alone occasion so late a stay. I watched therefore till the 24th of August, and then discovered that, under the eaves of the church, she attended upon two young, which were fledged, and now put out their white chins from a crevice. These remained till the twenty-seventh, looking more alert every day, and seeming to long to be on the wing. After this day they were missing at once; nor could I ever observe them with their dam coursing round the church in the act of learning to fly, as the first broods evidently do. On the thirty-first

I caused the eaves to be searched, but we found in the nest only two callow, dead, stinking swifts, on which a second nest had been formed. This double nest was full of the black shining cases of the *hippoboscae hirundinis.*

The following remarks on this unusual incident are obvious. The first is, that though it may be disagreeable to swifts to remain beyond the beginning of August, yet that they can subsist longer is undeniable. The second is, that this uncommon event, as it was owing to the loss of the first brood, so it corroborates my former remark, that swifts breed regularly but once; since, was the contrary the case, the occurrence above could neither be new nor rare.

PS – One swift was seen at Lyndon, in the country of Rutland, in 1782, so late as the third of September.

Gilbert White's tomb

Letter 53 also to the Hon. Daines Barrington

As I have sometimes known you make inquiries about several kinds of insects, I shall here send you an account of one sort which I little expected to have found in this kingdom. I had often observed that one particular part of a vine growing on the walls of my house was covered in the autumn with a black dust-like appearance, on which the flies fed eagerly; and that the shoots and leaves thus affected did not thrive; nor did the fruit ripen. To this substance I applied my glasses; but could not discover that it had anything to do with animal life, as I at first expected: but, upon a closer examination behind the larger boughs, we were surprised to find that they were coated over with husky shells, from whose side proceeded a cotton-like substance, surrounding a multitude of eggs. This curious and uncommon production put me upon recollecting what I have heard and read concerning the *coccus vitis viniferae* of Linnaeus, which, in the south of Europe, infests many vines, and is an horrid and loathsome pest. As soon as I had turned to the accounts given of this insect, I saw at once that it swarmed on my vine; and did not appear to have been at all checked by the preceding winter, which had been uncommonly severe.

Not being then at all aware that it had anything to do with England, I was much inclined to think that it came from Gibraltar among the many boxes and packages of plants and birds which I had formerly received from thence; and especially as the vine infested grew immediately under my study window, where I usually kept my specimens. True it is that I had received nothing from thence for some years: but as insects, we know, are conveyed from one country to another in a very unexpected manner, and have a wonderful power of maintaining their existence till they fall into a nidus proper for their support and increase, I cannot but suspect still that these cocci came to me originally from Andalusia. Yet, all the while, candour obliges me to confess that Mr Lightfoot has written me word that he once, and but once, saw these insects on a vine at Weymouth in Dorsetshire; which, it is here to be observed, is a seaport town to which the coccus might be conveyed by shipping.

As many of my readers may possibly never have heard of this strange and unusual insect, I shall here transcribe a passage from a natural history of Gibraltar, written by the Reverend John White, late Vicar of Blackburn in Lancashire, but not yet published –

> In the year 1770 a vine, which grew on the east side of my house, and which had produced the finest crops of grapes for years past, was suddenly overspread on all the woody branches with large lumps of a white fibrous substance resembling spiders' webs, or rather raw cotton. It was of a very clammy quality, sticking fast to everything that touched it, and capable of being spun into long threads. At first I suspected it to be the product of spiders, but could find none. Nothing was to be seen connected with it but many brown oval husky shells, which by no means looked like insects but rather resembled bits of the dry bark of the vine. The tree had a plentiful crop of grapes set, when this pest appeared upon it; but the fruit was manifestly injured by this foul incumbrance. It remained all the summer, still increasing, and loaded the woody and bearing branches to a vast degree. I often pulled off great quantities by handfuls; but it was so slimy and tenacious that it could by no means be cleared. The grapes never filled to their natural perfection, but turned watery and vapid. Upon perusing the works afterwards of M. de Réaumur, I found this matter perfectly described and accounted for. Those husky shells which I had observed, were no other than the

female coccus, from whose side this cotton-like substance exudes, and serves as a covering and security for their eggs.

To this account I think proper to add, that, though the female cocci are stationary, and seldom remove from the place to which they stick, yet the male is a winged insect; and that the black dust which I saw was undoubtedly the excrement of the females, which is eaten by ants as well as flies. Though the utmost severity of our winter did not destroy these insects, yet the attention of the gardener in a summer or two has entirely relieved my vinc from this filthy annoyance.

As we have remarked above that insects are often conveyed from one country to another in a very unaccountable manner, I shall here mention an emigration of small aphides, which was observed in the village of Selborne no longer ago than August the first, 1785.

About three o'clock in the afternoon of that day, which was very hot, the people of this village were surprised by a shower of aphides, or smother-flies, which fell in these parts. Those that were walking in the street at that juncture found themselves covered with these insects, which settled also on the hedges and gardens, blackening all the vegetables where they alighted. My annuals were discoloured with them, and the stalks of a bed of onions were quite coated over for six days after. These armies were then, no doubt, in a state of emigration, and shifting their quarters; and might have come, as far as we know, from the great hop-plantations of Kent or Sussex, the wind being all that day in the easterly quarter.[1] They were observed at the same time in great clouds about Farnham, and all along the vale from Farnham to Alton.[2]

1 For various methods by which several insects shift their quarters, see Derham's *Physico-Theology*.

2 Winged aphides are specially produced under certain circumstances as foundresses of colonies, and blow away at times in great numbers . ED.

Wood-pigeon

Letter 54 also to the Hon. Daines Barrington

Dear Sir – When I happen to visit a family where gold and silver fishes are kept in a glass bowl, I am always pleased with the occurrence, because it offers me an opportunity of observing the actions and propensities of those beings with whom we can be little acquainted in their natural state. Not long since I spent a fortnight at the house of a friend where there was such a vivary,[1] to which I paid no small attention, taking every occasion to remark what passed within its narrow limits. It was here that I first observed the manner in which fishes die. As soon as the creature sickens, the head sinks lower and lower, and it stands as it were on its head; till, getting weaker, and losing all poise, the tail turns over, and at last it floats on the surface of the water with its belly uppermost. The reason why fishes, when dead, swim in that manner is very obvious; because, when the body is no longer balanced by the fins of the belly, the broad muscular back preponderates by its own gravity, and turns the belly uppermost, as lighter from its being a cavity, and because it contains the swimming-bladders, which contribute to render it buoyant.

1 A rare and interesting case of a Latin word at one time thoroughly Englished and now once more employed with its classical termination. ED.

Some that delight in gold and silver fishes have adopted a notion that they need no aliment. True it is that they will subsist for a long time without any apparent food but what they can collect from pure water frequently changed; yet they must draw some support from animalcula, and other nourishment supplied by the water; because, though they seem to eat nothing, yet the consequences of eating often drop from them. That they are best pleased with such *jejune* diet may easily be confuted, since if you toss them crumbs they will seize them with great readiness, not to say greediness; however, bread should be given sparingly, lest, turning sour, it corrupt the water. They also feed on the water-plant called *Lemna* (ducks' meat), and also on small fry.

When they want to move a little, they gently protrude themselves with their *Pinnae pectorales*; but it is with their strong muscular tails only that they and all fishes shoot along with such inconceivable rapidity. It has been said that the eyes of fishes are immovable; but these apparently turn them forward or backward in their sockets as occasions require. They take little notice of a lighted candle, though applied close to their heads, but flounce and seem much frightened by a sudden stroke of the hand against the support whereon the bowl is hung; especially when they have been motionless, and are perhaps asleep. As fishes have no eye-lids, it is not easy to discern when they are sleeping or not, because their eyes are always open.

Nothing can be more amusing than a glass bowl containing such fishes; the double refractions of the glass and water represent them, when moving, in a shifting and changeable variety of dimensions, shades, and colours; while the two mediums, assisted by the concavo-convex shape of the vessel, magnify and distort them vastly; not to mention that the introduction of another element and its inhabitants into our parlours engages the fancy in a very agreeable manner.

Gold and silver fishes, though originally native of China and Japan, yet are become so well reconciled to our climate as to thrive and multiply very fast in our ponds and stews. Linnaeus ranks this species of fish, under the genus of *Cyprinus*, or carp, and calls it *Cyprinus auratus*.

Some people exhibit this sort of fish in a very fanciful way; for they cause a glass globe to be blown with a large hollow space within, that does not communicate with it. In this cavity they put a bird occasionally; so that you may see a goldfinch or a linnet hopping as it were in the midst of the water, and the fishes

The Yew Tree

swimming in a circle round it. The simple exhibition of the fishes is agreeable and pleasant; but in so complicated a way becomes whimsical and unnatural, and liable to the objection due to him,

Qui variare cupit rem prodigialiter unam.

I am, &c.

Letter 55 also to the Hon. Daines Barrington

October 10th, 1781

Dear Sir – I think I have observed before that much of the most considerable part of the house martins withdraw from hence about the first week in October; but that some, the latter broods I am now convinced, linger on till towards the middle of that month; and that at times, once perhaps in two or three years, a flight, for one day only, has shown itself in the first week in November.[1]

Having taken notice in October, 1780, that the last flight was numerous, amounting perhaps to one hundred and fifty; and that the season was soft and still; I was resolved to pay uncommon attention to these late birds; to find, if possible, where they roosted, and to determine the precise time of their retreat. The mode of life of these latter hirundines is very favourable to such a design; for they spend the whole day in the sheltered district between me and

1 The hirundines possessed a peculiar and fatal fascination for White. The suggestion at the end of this letter is one more contribution of a straw to his favourite mare's nest. ED.

the Hanger, sailing about in a placid, easy manner, and feasting on those insects which love to haunt a spot so secure from ruffling winds. As my principal object was to discover the place of their roosting, I took care to wait on them before they retired to rest, and was much pleased to find that for several evenings together, just at a quarter past five in the afternoon, they all scudded away in great haste towards the south-east, and darted down among the low shrubs above the cottages at the end of the hill. This spot in many respects seemed to be well calculated for their winter residence; for in many parts it is as steep as the roof of any house, and therefore secure from the annoyances of water; and it is moreover clothed with beechen shrubs, which, being stunted and bitten by sheep, make the thickest covert imaginable; and are so entangled as to be impervious to the smallest spaniel; besides it is the nature of underwood beech never to cast its leaf all the winter; so that, with the leaves on the ground and those on the twigs, no shelter can be more complete. I watched them on the thirteenth and fourteenth of October, and found their evening retreat was exact and uniform; but after this they made no regular appearance. Now and then a straggler was seen; and on the twenty-second of October, I observed two in the morning over the village, and with them my remarks for the season ended.

From all these circumstances put together, it is more than probable that this lingering flight, at so late a season of the year, never departed from the island. Had they indulged me that autumn with a November visit, as I much desired, I presume that, with proper assistants, I should have settled the matter past all doubt; but though the 3rd of November was a sweet day, and in appearance exactly suited to my wishes, yet not a martin was to be seen; and so I was forced, reluctantly, to give up the pursuit.

I have only to add that were the bushes, which cover some acres, and are not my own property, to be grubbed and carefully examined, probably those late broods, and perhaps the whole aggregate body of the house-martins of this district, might be found there, in different secret dormitories; and that, so far from withdrawing into warmer climes, it would appear that they never depart three hundred yards from the village.

Nuthatch

Letter 56 also to the Hon. Daines Barrington

They who write on natural history cannot too frequently advert to instinct, that wonderful limited faculty, which, in some instances, raises the brute creation as it were, above reason, and in others leaves them so far below it. Philosophers have defined instinct to be that secret influence by which every species is impelled naturally to pursue, at all times, the same way or track, without any teaching or example; whereas reason, without instruction, would often vary and do that by many methods which instinct effects by one alone. Now this maxim must be taken in a qualified sense; for there are instances in which instinct does vary and conform to the circumstances of place and convenience.

It has been remarked that every species of bird has a mode of nidification peculiar to itself, so that a schoolboy would at once pronounce on the sort of nest before him. This is the case among fields and woods, and wilds; but, in the villages round London, where mosses and gossamer, and cotton from vegetables, are hardly to be found, the nest of the chaffinch has not that elegant finished appearance, nor is it so beautifully studded with lichens, as in a more rural district; and the wren is obliged to construct its house with straws and dry grasses, which do not give it that rotundity and compactness so remarkable in the edifices of that little architect. Again, the regular nest of the house-martin is hemispheric; but where a rafter, or a joist, or a cornice, may happen to stand in the way, the nest is so contrived as to conform to the obstruction, and becomes flat, or compressed.

In the following instances instinct is perfectly uniform and consistent. There are three creatures, the squirrel, the field-mouse, and the bird called the nut-hatch (*sitta Europaea*), which live much on hazel-nuts; and yet they open them each in a different way. The first, after rasping off the small end, splits the shell in two with his long fore-teeth, as a man does with his knife; the second nibbles a hole with his teeth, so regular as if drilled with a wimble, and yet so small that one would wonder how the kernel can be extracted through it; while the last picks an irregular ragged hole with its bill; but as this artist has no paws to hold the nut firm while he pierces it, like an adroit workman, he fixes it, as it were in a vice, in some cleft of a tree, or in some crevice; when, standing over it, he perforates the stubborn shell. We have often placed nuts in the chink of a gate-post where nut-hatches have been known to haunt, and have always found that those birds have readily penetrated them. While at work they make a rapping noise that may be heard at a considerable distance.

You that understand both the theory and practical part of music may best inform us why harmony or melody should so strangely affect some men, as it were by recollection, for days after the concert is over. What I mean the following passage will most readily explain –

> *Praehabebat porro vocibus humanis, instrumentisque harmonicis musicam illam avium: non quod alia quoque non delectaretur: sed quod ex musica humana relinqueretur in animo continens quaedam, attentionemque et somnum conturbans agitatio; dum ascensus, exscensus, tenores, ac mutationes illae sonorum et consonantiarum euntque, redeuntque per phantasiam – cum nihil tale relinqui possit ex modulationibus avium, quae, quod non sunt perinde a nobis imitabiles, non possunt perinde internam facultatem commovere.*
>
> GASSENDUS in *Vita Peireskii*

This curious quotation strikes me much by so well representing my own case, and by describing what I have so often felt, but never could so well express. When I hear fine music I am haunted with

The Plestor and Hanger

passages therefrom night and day; and especially at first waking, which by their importunity, give me more uneasiness than pleasure; elegant lessons still tease my imagination, and recur irresistibly to my recollection even at seasons when I am desirous of thinking of more serious matters.

I am, &c.

Peregrine Falcon

Letter 57 also to the Hon. Daines Barrington

A rare, and I think a new, little bird frequents my garden, which I have great reason to think is the pettichaps: it is common in some parts of the kingdom; and I have received formerly several dead specimens from Gibraltar. This bird much resembles the white-throat, but has a more white or rather silvery breast and belly; is restless and active, like the willow-wrens, and hops from bough to bough, examining every part for food; it also runs up the stems of the crown-imperials, and, putting its head into the bells of those flowers, sips the liquor which stands in the nectarium of each petal. Sometimes it feeds on the ground like the hedge-sparrow, by hopping about on the grass-plots and mown walks.

One of my neighbours, an intelligent and observing man, informs me that, in the beginning of May, and about ten minutes before eight o'clock in the evening, he discovered a great cluster of house-swallows, thirty, at least, he supposes, perching on a willow that hung over the verge of James Knight's upper pond. His attention was first drawn by the twittering of these birds, which sat motionless in a row on the bough, with their heads all one way, and, by their weight, pressing down the twig so that it nearly touched the water. In this situation he watched them till he could see no longer. Repeated accounts of this sort, spring and fall, induce us greatly to suspect that house-swallows have some strong attachment to water, independent of the matter of food; and, though they may not retire into that element, yet they may conceal themselves in the banks of pools and rivers during the uncomfortable months of winter.

One of the keepers of Wolmer Forest sent me a peregrine falcon, which he shot on the verge of that district as it was devouring a wood-pigeon. The *falco peregrinus*, or haggard-falcon, is a noble species of hawk seldom seen in the southern counties. In winter 1767, one was killed in the neighbouring parish of Faringdon, and sent by me to Mr Pennant into North Wales.[1] Since that time I have met with none till now. The specimen mentioned above was in fine preservation, and not injured by the shot: it measured forty-two inches from wing to wing, and twenty-one from beak to tail, and weighed two pounds and an half standing weight. This species is very robust, and wonderfully formed for rapine; its breast was plump and muscular; its thighs long, thick, and brawny; and its legs remarkably short and well set: the feet were armed with most formidable, sharp, long talons: the eyelids and cere of the bill were yellow: but the irides of the eyes dusky; the beak was thick and hooked, and of a dark colour, and had a jagged process near the end of the upper mandible on each side: its tail, or train, was short in proportion to the bulk of its body; yet the wings, when closed, did not extend to the end of the train. From its large and fair proportions it might be supposed to have been a female; but I was not permitted to cut open the specimen. For one of the birds of prey, which are usually lean, this was in high case: in its craw were many barley-corns, which probably came from the crop of the wood-pigeon, on which it was feeding when shot; for voracious birds do not eat grain, but when devouring their quarry, with undistinguishing vehemence swallow bones and feathers, and all matters, indiscriminately. This falcon was probably driven from the mountains of North Wales or Scotland, where they are known to breed, by rigorous weather and deep snows that had lately fallen.

I am, &c.

1 See my tenth and eleventh letter to that gentleman.

Hen-harrier

Letter 58 also to the Hon. Daines Barrington

My near neighbour, a young gentleman in the service of the East India Company, has brought home a dog and a bitch of the Chinese breed from Canton, such as are fattened in that country for the purpose of being eaten: they are about the size of a moderate spaniel; of a pale yellow colour, with coarse bristling hairs on their backs; sharp upright ears, and peaked heads, which give them a very foxlike appearance. Their hind legs are unusually straight, without any bend at the hock or ham, to such a degree as to give them an awkward gait when they trot. When they are in motion their tails are curved high over their backs like those of some hounds, and have a bare place each on the outside from the tip midway, that does not seem to be matter of accident, but somewhat singular. Their eyes are jet-black, small, and piercing; the insides of their lips and mouths of the same colour, and their tongues blue. The bitch has a dew-claw on each hind leg; the dog has none. When taken out into a field the bitch showed some disposition for hunting, and dwelt on the scent of a covey of partridges till she sprung them, giving her tongue all the time. The dogs in South America are dumb; but these bark much in a short thick manner like foxes, and have a surly, savage demeanour like their ancestors, which are not domesticated, but bred up in sties, where they are fed for the table with rice-meal and other farinaceous food. These dogs, having been taken on board as soon as weaned, could not learn much from their dam; yet they did not relish flesh when they came to England. In the islands of the Pacific ocean the dogs are bred up on vegetables, and would not eat flesh when offered them by our circumnavigators.

We believe that all dogs, in a state of nature, have sharp, upright, foxlike ears; and that hanging ears, which are esteemed so graceful, are the effect of choice breeding and cultivation. Thus, in the *Travels of Ysbrandt Ides from Muscovy to China*, the dogs which draw the Tartars on snow-sledges, near the river Oby, are engraved with prick-ears, like those from Canton. The Kamschatdales also train the same sort of sharp-eared, peak-nosed dogs to draw their sledges; as may be seen in an elegant print engraved for Captain Cook's last voyage round the world.

Now we are upon the subject of dogs, it may not be impertinent to add, that spaniels, as all sportsmen know, though they hunt partridges and pheasants as it were by instinct, and with much delight and alacrity, yet will hardly touch their bones when offered as food; nor will a mongrel dog of my own, though he is remarkable for finding that sort of game. But when we came to offer the bones of partridges to the two Chinese dogs, they devoured them with much greediness, and licked the platter clean.

No sporting dogs will flush woodcocks till inured to the scent and trained to the sport, which they then pursue with vehemence and transport; but then they will not touch their bones, but turn from them with abhorrence, even when they are hungry.

Now, that dogs should not be fond of the bones of such birds as they are not disposed to hunt is no wonder; but why they reject and do not care to eat their natural game is not so easily accounted for, since the end of hunting seems to be, that the chase pursued should be eaten. Dogs again will not devour the more rancid water-fowls, nor indeed the bones of any wild fowls; nor will they touch the foetid bodies of birds that feed on offal and garbage; and indeed there may be somewhat of providential instinct in this circumstance of dislike; for vultures,[1] and kites, and ravens, and crows, &c., were intended to be messmates with dogs[2] over their carrion; and seem to be appointed by Nature as fellow-scavengers to remove all cadaverous nuisances from the face of the earth.

I am, &c.

1 Hasselquist, in his *Travels to the Levant*, observes that the dogs and vultures at Grand Cairo maintain such a friendly intercourse as to bring up their young together in the same place.

2 The Chinese word for a dog to an European ear sounds like *quihloh*.

Lyss old church

White's Thrush

Letter 59

The fossil wood buried in the bogs of Wolmer Forest is not yet all exhausted; for the peat-cutters now and then stumble upon a log. I have just seen a piece which was sent by a labourer of Oakhanger to a carpenter of this village; this was the butt-end of a small oak, about five feet long, and about five inches in diameter. It had apparently been severed from the ground by an axe, was very ponderous, and as black as ebony. Upon asking the carpenter for what purpose he had procured it, he told me that it was to be sent to his brother, a joiner at Farnham, who was to make use of it in cabinet-work, by inlaying it along with whiter woods.

Those that are much abroad on evenings after it is dark, in spring and summer, frequently hear a nocturnal bird passing by on the wing, and repeating often a short, quick note. This bird I have remarked myself, but never could make out till lately. I am assured now that it is the stone-curlew (*charadrius oedicnemus*). Some of them pass over or near my house almost every evening after it is dark, from the uplands of the hill and North Fields, away down towards Dorton, where, among the streams and meadows, they find a greater plenty of food. Birds that fly by night are obliged to be noisy; their notes often repeated become signals or watch-words to keep them together, that they may not stray or lose each the other in the dark.

The evening proceedings and manoeuvres of the rooks are curious and amusing in the autumn. Just before dusk they return in long strings from the foraging of the day, and rendezvous by thousands over Selborne Down, where they wheel round in the air and sport and dive in a playful manner, all the while exerting their voices, and making a loud cawing, which, being blended and softened by the distance that we at the village are below them, becomes a confused noise or chiding; or rather a pleasing murmur, very engaging to the imagination, and not unlike the cry of a pack of hounds in hollow, echoing woods, or the rushing of the wind in tall trees, or the tumbling of the tide upon a pebbly shore. When this ceremony is over, with the last gleam of day, they retire for the night to the deep beechen woods of Tisted and Ropley. We remember a little girl who, as she was going to bed, used to remark on such an occurrence, in the true spirit of physico-theology, that the rooks were saying their prayers; and yet this child was much too young to be aware that the Scriptures have said of the Deity – that 'he feedeth the ravens who call upon him'.

I am, &c.

Letter 60 also to the Hon. Daines Barrington

In reading Dr Huxam's *Observationes de Aëre*, &c., written at Plymouth, I find by those curious and accurate remarks, which contain an account of the weather from the year 1727 to the year 1748 inclusive, that though there is frequent rain in that district of Devonshire, yet the quantity falling is not great; and that some years it has been very small: for in 1731 the rain measured only 17.266 in.; and in 1741, 20.354 in.; and again, in 1743, only 20.908 in. Places near the sea have frequent scuds, that keep the atmosphere moist, yet do not reach far up into the country; making thus the maritime situations appear wet, when the rain is not considerable. In the wettest years at Plymouth the doctor measured only once 36; and again once, viz. 1734, 37.114 in. – a quantity of rain that has twice been exceeded at Selborne in the short period of my observations. Dr Huxam remarks that frequent small rains keep the air moist; while heavy ones render it more dry, by beating down the vapours. He is also of opinion that the dingy, smoky appearance in the sky, in very dry seasons, arises from the want of moisture sufficient to let the light through, and render the atmosphere transparent; because he had observed several bodies more diaphanous when wet than dry, and did never recollect that the air had that look in rainy seasons.

My friend, who lives just beyond the top of the down, brought his three swivel guns to try them in my outlet, with their muzzles towards the Hanger, supposing that the report would have had a great effect; but the experiment did not answer his expectation. He then removed them to the alcove on the Hanger; when the sound, rushing along the Lythe and Comb-wood was very grand; but it was at the Hermitage that the echoes and repercussions delighted the hearers; not only filling the Lythe with the roar, as if all the beeches were tearing up by the roots; but, turning to the left, they pervaded the vale above Comb-wood ponds, and after a pause seemed to take up the crash again, and to extend round Hartley Hangers, and to die away at last among the coppices and coverts of Ward-le-Ham. It has been remarked before that this district is an Anathoth, a place of responses or echoes, and therefore proper for such experiments: we may farther add that the pauses in echoes, when they cease and yet are taken up again, like the pauses in music, surprise the hearers, and have a fine effect on the imagination.

The gentleman above-mentioned has just fixed a barometer[1] in his parlour at Newton Valence. The tube was first filled here (at Selborne) twice with care, when the mercury agreed and stood exactly with my own; but, being filled twice again at Newton, the mercury stood, on account of the great elevation of that house, three-tenths of an inch lower than the barometers at this village, and so continues to do, be the weight of the atmosphere what it may. The plate of the barometer at Newton is figured as low as 27; because in stormy weather the mercury there will sometimes descend below 28. We have supposed Newton House to stand two hundred feet higher than this house: but if the rule holds good, which says that mercury in a barometer sinks one-tenth of an inch for every hundred feet elevation, then the Newton barometer, by standing three-tenths lower than that of Selborne, proves that Newton House must be three hundred feet higher than that in which I am writing, instead of two hundred.

1 This barometer can still be seen at Newton Valence vicarage. The incumbent at this time and for many years after was the Revd Edmund White, Gilbert White's nephew. E.H.N.

It may not be impertinent to add, that the barometers at Selborne stand three-tenths of an inch lower than the barometers at South Lambeth: whence we may conclude that the former place is about three hundred feet higher than the latter; and with good reason, because the streams that rise with us run into the Thames at Weybridge, and so to London. Of course, therefore, there must be lower ground all the way from Selborne to South Lambeth; the distance between which, all the windings and indentings of the streams considered, cannot be less than an hundred miles.

I am, &c

Faringdon Church

The Wakes

Letter 61 also to the Hon. Daines Barrington

Since the weather of a district is undoubtedly part of its natural history, I shall make no further apology for the four following letters, which will contain many particulars concerning some of the great frosts, and a few respecting some very hot summers, that have distinguished themselves from the rest during the course of my observations.

As the frost of January 1768 was, for the small time it lasted, the most severe that we had then known for many years, and was remarkably injurious to evergreens, some account of its rigour, and reason of its ravages, may be useful, and not unacceptable to persons that delight in planting and ornamenting; and may particularly become a work that professes never to lose sight of utility.[1]

For the last two or three days of the former year there were considerable falls of snow, which lay deep and uniform on the

1 This is the first overt indication White has given of the deliberate intention to write a book. ED

ground without any drifting, wrapping up the more humble vegetation in perfect security. From the first day to the fifth of the new year more snow succeeded; but from that day the air became entirely clear, and the heat of the sun about noon had a considerable influence in sheltered situations.

It was in such an aspect that the snow on the author's evergreens[2] was melted every day, and frozen intensely every night; so that the laurustines, bays, laurels, and arbutuses looked, in three or four days, as if they had been burnt in the fire; while a neighbour's plantation of the same kind, in a high cold situation, where the snow was never melted at all, remained uninjured.

From hence I would infer that it is the repeated melting and freezing of the snow that is so fatal to vegetation, rather than the severity of the cold.[3] Therefore it highly behoves every planter, who wishes to escape the cruel mortification of losing in a few days the labour and hopes of years, to bestir himself on such emergencies; and if his plantations are small, to avail himself of mats, cloths, pease-haulm, straw, reeds, or any such covering, for a short time; or, if his shrubberies are extensive, to see that his people go about with prongs and forks, and carefully dislodge the snow from the boughs: since the naked foliage will shift much better for itself, than where the snow is partly melted and frozen again.

It may perhaps appear at first like a paradox; but doubtless the more tender trees and shrubs should never be planted in hot aspects; not only for the reason assigned above, but also because, thus circumstanced, they are disposed to shoot earlier in the spring, and to grow on later in the autumn than they would otherwise do, and so are sufferers by lagging or early frosts. For this reason also plants from Siberia will hardly endure our climate; because, on the very first advances of spring, they shoot away, and so are cut off by the severe nights of March or April.

Dr Fothergill and others have experienced the same inconvenience with respect to the more tender shrubs from North America, which they therefore plant under north walls. There should also,

2 The phrase 'the author', which occurs here and in some subsequent passages, indicates the unreality of these later letters. ED

3 This observation has since been abundantly justified. I have myself observed that near the summit of Hind Head in Surrey, over eight hundred feet in height, many trees and shrubs pass uninjured through severe winters, while below seven hundred feet, on the same hill, many individuals of identical species are destroyed by the repeated thawings and freezings. ED.

perhaps, be a wall to the east to defend them from the piercing blasts from that quarter.

This observation might without any impropriety be carried into animal life; for discerning bee-masters now find that their hives should not in the winter be exposed to the hot sun, because such unseasonable warmth awakens the inhabitants too early from their slumbers; and by putting their juices into motion too soon, subjects them afterwards to inconveniences when rigorous weather returns.

The coincidents attending this short but intense frost were, that the horses fell sick with an epidemic distemper, which injured the winds of many, and killed some; that colds and coughs were general among the human species; that it froze under people's beds for several nights; that meat was so hard frozen that it could not be spitted, and could not be secured but in cellars; that several redwings and thrushes were killed by the frost; and that the large titmouse continued to pull straws lengthwise from the eaves of thatched houses and barns in a most adroit manner for a purpose that has been explained already.[4]

On the 3rd of January, Benjamin Martin's thermometer[5] within doors, in a close parlour where there was no fire, fell in the night to 20°, and on the 4th, to 18°, and on the 7th, to 17½°, a degree of cold which the owner never since saw in the same situation; and he regrets much that he was not able at that juncture to attend his instrument abroad. All this time the wind continued north and north-east; and yet on the 8th roost-cocks, which had been silent, began to sound their clarions, and crows to clamour, as prognostic of milder weather; and, moreover, moles began to heave and work, and a manifest thaw took place. From the latter circumstance we may conclude that thaws often originate under ground from warm vapours which arise; else how should subterraneous animals receive such early intimations of their approach. Moreover, we have often observed that cold seems to descend from above; for when a thermometer hangs abroad in a frosty night, the intervention of a cloud shall immediately raise the mercury 10°; and a clear sky shall again compel it to descend to its former gage.[6]

And here it may be proper to observe, on what has been said above, that though frosts advance to their utmost severity by

4 See Letter 41 to Mr Pennant.

5 Benjamin Martin was a maker of scientific instruments. ED.

6 This is a first indication of the importance of radiation, and of the value of clouds as an earth-blanket, since so fully worked out by Tyndall. ED.

somewhat of a regular gradation, yet thaws do not usually come on by as regular a declension of cold, but often take place immediately from intense freezing; as men in sickness often mend at once from a paroxysm.

To the great credit of Portugal laurels and American junipers, be it remembered that they remained untouched amidst the general havoc: hence men should learn to ornament chiefly with such trees as are able to withstand accidental severities, and not subject themselves to the vexation of a loss which may befall them once perhaps in ten years, yet may hardly be recovered through thc whole course of their lives.

As it appeared afterwards, the ilexes were much injured, the cypresses were half destroyed, the arbutuses lingered on, but never recovered; and the bays, laurustines, and laurels, were killed to the ground; and the very wild hollies, in hot aspects, were so much affected that they cast all their leaves.

By the 14th of January the snow was entirely gone; the turnips emerged not damaged at all, save in sunny places; the wheat looked delicately, and the garden plants were well preserved; for snow is the most kindly mantle that infant vegetation can be wrapped in: were it not for that friendly meteor no vegetable life could exist at all in northerly regions. Yet in Sweden the earth in April is not divested of snow for more than a fortnight before the face of the country is covered with flowers.

Great Northern Diver

Letter 62 also to the Hon. Daines Barrington

There were some circumstances attending the remarkable frost in January, 1776, so singular and striking, that a short detail of them may not be unacceptable.

The most certain way to be exact will be to copy the passages from my journal, which were taken from time to time, as things occurred. But it may be proper previously to remark that the first week in January was uncommonly wet, and drowned with vast rains from every quarter: from whence may be inferred, as there is great reason to believe is the case, that intense frosts seldom take place till the earth is perfectly glutted and chilled with water;[1] and hence dry autumns are seldom followed by rigorous winters.

January 7th – Snow driving all the day, which was followed by frost, sleet, and some snow, till the 12th, when a prodigious mass overwhelmed all the works of men, drifting over the tops of the gates and filling the hollow lanes.

On the 14th the writer was obliged to be much abroad; and thinks

1 The autumn preceding January 1768 was very wet, and particularly the month of September, during which there fell at Lyndon, in the county of Rutland, six inches and a half of rain. And the terrible long frost in 1739–40 set in after a rainy season, and when the springs were very high.

he never before or since has encountered such rugged Siberian weather. Many of the narrow roads were now filled above the tops of the hedges; through which the snow was driven into most romantic and grotesque shapes, so striking to the imagination as not to be seen without wonder and pleasure. The poultry dared not to stir out of their roosting-places; for cocks and hens are so dazzled and confounded by the glare of snow that they would soon perish without assistance. The hares also lay sullenly in their seats, and would not move till compelled by hunger; being conscious – poor animals – that the drifts and heaps treacherously betray their footsteps, and prove fatal to numbers of them.

From the 14th the snow continued to increase, and began to stop the road wagons and coaches, which could no longer keep on their regular stages; and especially on the western roads, where the fall appears to have been deeper than in the south. The company at Bath, that wanted to attend the Queen's birthday, were strangely incommoded: many carriages of persons who got in their way to town from Bath as far as Marlborough, after strange embarrassments, here met with a *ne plus ultra.* The ladies fretted, and offered large rewards to labourers if they would shovel them a track to London; but the relentless heaps of snow were too bulky to be removed; and so the 18th passed over, leaving the company in very uncomfortable circumstances at the Castle and other inns.

On the 20th the sun shone out for the first time since the frost began; a circumstance that has been remarked before much in favour of vegetation. All this time the cold was not very intense, for the thermometer stood at 29°, 28°, 25°, and thereabout; but on the 21st it descended to 20°. The birds now began to be in a very pitiable and starving condition. Tamed by the season, skylarks settled in the streets of towns, because they saw the ground was bare; rooks frequented dunghills close to houses; and crows watched horses as they passed, and greedily devoured what dropped from them; hares now came into men's gardens, and, scraping away the snow, devoured such plants as they could find.

On the 22nd the author had occasion to go to London through a sort of Laplandian scene, very wild and grotesque indeed. But the metropolis itself exhibited a still more singular appearance than the country; for being bedded deep in snow, the pavement of the streets could not be touched by the wheels or the horses' feet, so that the carriages ran about without the least noise. Such an exemption from din and clatter was strange, but not pleasant; it seemed to convey an uncomfortable idea of desolation –

Temple

. . . Ipsa silentia terrent.

On the 27th much snow fell all day, and in the evening the frost became very intense. At South Lambeth, for the four following nights, the thermometer fell to 11°, 7°, 6°, 6°, and at Selborne to 7°, 6°, 10°, and on the 31st of January, just before sunrise, with rime on the trees and on the tube of the glass, the quicksilver sunk exactly to zero, being 32° below the freezing point; but by eleven in the morning, though in the shade, it sprang up to 16½° – a most unusual degree of cold this for the south of England![2] During these four nights the cold was so penetrating that it occasioned ice in warm chambers and under beds; and in the day the wind was so keen that persons of robust constitutions could scarcely endure to face it. The Thames was at once so frozen over both above and

2 At Selborne the cold was greater than at any other place that the author could hear of with certainty: though some reported at the time that at a village in Kent the thermometer fell two degrees below zero, viz., thirty-four degrees below the freezing point.

The thermometer used at Selborne was graduated by Benjamin Martin.

below bridge that crowds ran about on the ice. The streets were now strangely encumbered with snow, which crumbled and trod dusty; and, turning grey, resembled bay-salt; what had fallen on the roofs was so perfectly dry that, from first to last, it lay twenty-six days on the houses in the city; a longer time than had been remembered by the oldest housekeepers living. According to all appearances we might now have expected the continuance of this rigorous weather for weeks to come, since every night increased in severity; but, behold, without any apparent cause, on the 1st of February a thaw took place, and some rain followed before night, making good the observation above, that frosts often go off as it were at once, without any gradual declension of cold. On the 2nd of February the thaw persisted; and on the 3rd swarms of little insects were frisking and sporting in a court-yard at South Lambeth, as if they had felt no frost. Why the juices in the small bodies and smaller limbs of such minute beings are not frozen is a matter of curious inquiry.

Severe frosts seem to be partial, or to run in currents; for at the same juncture, as the author was informed by accurate correspondents, at Lyndon, in the county of Rutland, the thermometer stood at 19°; at Blackburn, in Lancashire, at 19°; and at Manchester at 21°, 20°, and 18°. Thus does some unknown circumstance strangely overbalance latitude, and render the cold sometimes much greater in the southern than the northern parts of this kingdom.

The consequences of this severity were, that in Hampshire, at the melting of the snow, the wheat looked well, and the turnips came forth little injured. The laurels and laurustines were somewhat damaged, but only in hot aspects. No evergreens were quite destroyed; and not half the damage sustained that befel in January 1768. Those laurels that were a little scorched on the south sides were perfectly untouched on their north sides. The care taken to shake the snow day by day from the branches seemed greatly to avail the author's evergreens. A neighbour's laurel-hedge, in a high situation, and facing to the north, was perfectly green and vigorous; and the Portugal laurels remained unhurt.

As to the birds, the thrushes and blackbirds were mostly destroyed; and the partridges, by the weather and poachers, were so thinned that few remained to breed the following year.

Letter 63 also to the Hon. Daines Barrington

As the frost in December 1784 was very extraordinary, you, I trust, will not be displeased to hear the particulars; and especially when I promise to say no more about the severities of winter after I have finished this letter.[1]

The first week in December was very wet, with the barometer very low. On the 7th, with the barometer at 28.5″ came on a vast snow, which continued all that day and the next, and most part of the following night; so that by the morning of the 9th the works of men were quite overwhelmed, the lanes filled so as to be impassable, and the ground covered twelve or fifteen inches without any drifting. In the evening of the 9th the air began to be so very sharp that we thought it would be curious to attend to the motions of a thermometer; we therefore hung out two, one made by Martin and one by Dollond, which soon began to show us what we were to expect; for by ten o'clock they fell to 21°, and at eleven to 4°, when we went to bed. On the 10th, in the morning, the quicksilver of

1 White here pretends to be still writing letters, but the pretence by this time has become sufficiently transparent. ED.

Dollond's glass was down to half a degree below zero; and that of Martin's, which was absurdly graduated only to four degrees above zero, sunk quite into the brass guard of the ball; so that when the weather became most interesting this was useless. On the 10th, at eleven at night, though the air was perfectly still, Dollond's glass went down to one degree below zero! This strange severity of the weather made me very desirous to know what degree of cold there might be in such an exalted and near situation as Newton. We had therefore, on the morning of the 10th, written to Mr —, and entreated him to hang out his thermometer, made by Adams, and to pay some attention to it morning and evening, expecting wonderful phenomena, in so elevated a region, at two hundred feet or more above my house. But, behold! on the 10th, at eleven at night, it was down only to 17°, and the next morning at 22°, when mine was at 10°! We were so disturbed at this unexpected reverse of comparative local cold, that we sent one of my glasses up, thinking that of Mr — must, somehow, be wrongly constructed. But, when the instruments came to be confronted, they went exactly together; so that for one night at least, the cold at Newton was 18° less than at Selborne; and, through the whole frost, 10° or 12°.[2] And indeed, when we came to observe consequences, we could readily credit this; for all my laurustines, bays, ilexes, arbutuses, cypresses, and even my Portugal laurels,[3] and (which occasions more regret) my fine sloping laurel-hedge, were scorched up; while at Newton, the same trees have not lost a leaf!

We had steady frost on to the 25th, when the thermometer in the morning was down to 10° with us, and at Newton only to 21°. Strong frost continued till the 31st, when some tendency to thaw was observed; and, by January the 3rd, 1785, the thaw was confirmed, and some rain fell.

A circumstance that I must not omit, because it was new to us, is, that on Friday, December the 10th, being bright sunshine, the air was full of icy *spiculae*, floating in all directions, like atoms in a sunbeam let into a dark room. We thought them at first particles of

2 I have observed an exactly similar fact on Hind Head, where the thermometer on frosty nights often stands higher than in the valley below. ED.

3 Mr Miller, in his *Gardener's Dictionary*, says positively that the Portugal laurels remained untouched in the remarkable frost of 1739–40. So that either that accurate observer was much mistaken, or else the frost of December 1784 was much more severe and destructive than that in the year above-mentioned.

the rime falling from my tall hedges; but were soon convinced to the contrary, by making our observations in open places where no rime could reach us. Were they watery particles of the air frozen as they floated, or were they evaporations from the snow frozen as they mounted?

We were much obliged to the thermometers for the early information they gave us; and hurried our apples, pears, onions, potatoes, &c., into the cellar, and warm closets; while those who had not, or neglected such warnings, lost all their store of roots and fruits, and had their very bread and cheese frozen.

I must not omit to tell you that, during these two Siberian days, my parlour cat was so electric, that had a person stroked her, and been properly insulated, the shock might have been given to a whole circle of people.

I forgot to mention before, that, during the two severe days, two men, who were tracing hares in the snow, had their feet frozen; and two men, who were much better employed, had their fingers so affected by the frost while they were thrashing in a barn, that a mortification followed, from which they did not recover for many weeks.

This frost killed all the furze and most of the ivy, and in many places stripped the hollies of all their leaves. It came at a very early time of the year, before old November ended; and yet may be allowed from its effects to have exceeded any since 1730–40.

Hybrid Bird

Letter 64 also to the Hon. Daines Barrington

As the effects of heat are seldom very remarkable in the northerly climate of England, where the summers are often so defective in warmth and sunshine as not to ripen the fruits of the earth so well as might be wished, I shall be more concise in my account of the severity of a summer season, and so make a little amends for the prolix account of the degrees of cold, and the inconveniences that we suffered from some late rigorous winters.

The summers of 1781 and 1783 were unusually hot and dry; to them therefore I shall turn back in my journals, without recurring to any more distant period. In the former of these years my peach and nectarine trees suffered so much from the heat that the rind on the bodies was scalded and came off; since which the trees have been in a decaying state. This may prove a hint to assiduous gardeners to fence and shelter their wall-trees with mats or boards, as they may easily do, because such annoyance is seldom of long continuance. During that summer also, I observed that my apples were coddled, as it were, on the trees; so that they had no quickness of flavour, and would not keep in the winter. This circumstance put me in mind of what I have heard travellers assert, that they never ate a good apple or apricot in the south of Europe, where the heats were so great as to render the juices vapid and insipid.

The great pests of a garden are wasps, which destroy all the finer fruits just as they are coming into perfection. In 1781 we had none; in 1783 there were myriads; which would have devoured all the produce of my garden, had not we set the boys to take the nests, and caught thousands with hazel-twigs tipped with birdlime: we have since employed the boys to take and destroy the large breeding wasps in the spring.[1] Such expedients have a great effect on these marauders, and will keep them under. Though wasps do not abound but in hot summers, yet they do not prevail in every hot summer, as I have instanced in the two years above-mentioned.

In the sultry season of 1783, honey-dews were so frequent as to deface and destroy the beauties of my garden. My honeysuckles, which were one week the most sweet and lovely objects that the eye could behold, became the next the most loathsome; being enveloped in a viscous substance, and loaded with black aphides, or smother-flies. The occasion of this clammy appearance seems to be this, that in hot weather the effluvia of flowers in fields and meadows and gardens are drawn up in the day by a brisk evaporation, and then in the night fall down again with the dews, in which they are entangled; that the air is strongly scented, and therefore impregnated with the particles of flowers in summer weather, our senses will inform us; and that this clammy sweet substance is of the vegetable kind we may learn from bees, to whom it is very grateful: and we may be assured that it falls in the night, because it is always first seen in warm still mornings.[2]

On chalky and sandy soils, and in the hot villages about London, the thermometer has been often observed to mount as high as 83° or 84°; but with us, in this hilly and woody district, I have hardly ever seen it exceed 80°; nor does it often arrive at that pitch. The reason, I conclude, is that our dense clayey soil, so much shaded by trees, is not so easily heated through as those above-mentioned: and, besides, our mountains cause currents of air and breezes; and the vast effluvia from our woodlands temper and moderate our heats.

1 These are what are known as 'foundress wasps' – impregnated queens which struggle through the winter and become mothers of colonies in the succeeding season. The destruction of one such pregnant female in early spring is equivalent to the destruction of an entire nest in summer. ED.

2 Honey-dew is now known to be mainly produced by aphides, which White here incidentally notices side by side with it, without suspecting their causal connection. It is possible that a small amount of honey-dew may be exuded by the plants themselves, but by far the greater portion is undoubtedly due to the secretions of plant-lice. ED.

Hollow Lane near Norton

Letter 65 also to the Hon. Daines Barrington

The summer of the year 1783 was an amazing and portentous one, and full of horrible phaenomena for, besides the alarming meteors and tremendous thunder-storms that affrighted and distressed the different counties of this kingdom, the peculiar haze, or smoky fog, that prevailed for many weeks in this island, and in every part of Europe, and even beyond its limits, was a most extraordinary appearance, unlike anything known within the memory of man. By my journal I find that I had noticed this strange occurrence from June 23rd to July 20th inclusive, during which period the wind varied to every quarter without making any alteration in the air. The sun, at noon, looked as blank as a clouded moon, and shed a rust-coloured ferruginous light on the ground, and floors of rooms; but was particularly lurid and blood-coloured at rising and setting![1]

1 The close resemblance of these phenomena to those which were observed to follow the great eruption of Krakatoa in Java renders it almost certain

All the time the heat was so intense that butchers' meat could hardly be eaten on the day after it was killed; and the flies swarmed so in the lanes and hedges that they rendered the horses half frantic, and riding irksome. The country people began to look with a superstitious awe at the red, louring aspect of the sun; and indeed there was reason for the most enlightened person to be apprehensive; for, all the while, Calabria and part of the isle of Sicily, were torn and convulsed with earthquakes; and about that juncture a volcano sprang out of the sea on the coast of Norway. On this occasion Milton's noble simile of the sun, in his first book of *Paradise Lost*, frequently occurred to my mind; and it is indeed particularly applicable, because, towards the end, it alludes to a superstitious kind of dread, with which the minds of men are always impressed by such strange and unusual phenomena.

> . . . As when the sun, new risen,
> Looks through the horizontal, misty air,
> Shorn of his beams; or from behind the moon,
> In dim eclipse, disastrous twilight sheds
> On half the nations, and with fear of change
> Perplexes monarchs.

---

that they were due to a similar volcanic origin. This is the more likely since White specially notices volcanic activity throughout Europe as concomitants of the lurid sunsets. But the volcanic dust on which these appearances doubtless depended may more likely have come from some extra-European crater, whose activity coincided with that of the European system. ED.

Grange Farm

Letter 66 also to the Hon. Daines Barrington

We are very seldom annoyed with thunder-storms; and it is no less remarkable than true, that those which arise in the south have hardly been known to reach this village; for, before they get over us, they take a direction to the east or to the west, or sometimes divide in two, go in part to one of those quarters, and in part to the other; as was truly the case in summer 1783, when, though the country round was continually harassed with tempests, and often from the south, yet we escaped them all, as appears by my journal of that summer. The only way that I can at all account for this fact – for such it is – is that, on that quarter, between us and the sea, there are continual mountains, hill behind hill, such as Nore-hill, the Barnet, Butser-hill, and Portsdown, which somehow divert the storms, and give them a different direction. High promontories, and elevated grounds, have always been observed to attract clouds and disarm them of their mischievous contents, which are discharged into the trees and summits as soon as they come in contact with those turbulent meteors; while the humble vales escape, because they are so far beneath them.

But, when I say I do not remember a thunder-storm from the south, I do not mean that we never have suffered from thunder-storms at all; for on June 5th, 1784, the thermometer in the morning being 64°, and at noon 70°, the barometer at 29.65″, and the wind north, I observed a blue mist, smelling strongly of sulphur, hanging along our sloping woods, and seeming to indicate that thunder was at hand. I was called in about two in the afternoon, and so missed seeing the gathering of the clouds in the north; which they who were abroad assured me had something uncommon in its appearance. At about a quarter after two the storm began in the parish of Hartley, moving slowly from north to south; and from thence it came over Norton-farm, and so to Grange-farm, both in this parish. It began with vast drops of rain, which were soon succeeded by round hail, and then by convex pieces of ice, which measured three inches in girth. Had it been as extensive as it was violent, and of any continuance (for it was very short), it must have ravaged all the neighbourhood. In the parish of Hartley it did some damage to one farm; but Norton, which lay in the centre of the storm, was greatly injured; as was Grange, which lay next to it. It did but just reach to the middle of the village, where the hail broke my north windows, and all my garden lights and hand-glasses, and many of my neighbours' windows. The extent of the storm was about two miles in length and one in breadth. We were just sitting down to dinner; but were soon diverted from our repast by the clattering of tiles and the jingling of glass. There fell at the same time prodigious torrents of rain on the farms above-mentioned, which occasioned a flood as violent as it was sudden; doing great damage to the meadows and fallows, by deluging the one and washing away the soil of the other. The hollow lane towards Alton was so torn and disordered as not to be passable till mended, rocks being removed that weighed two hundred weight. Those that saw the effect which the great hail had on ponds and pools say that the dashing of the water made an extraordinary appearance, the froth and spray standing up in the air three feet above the surface. The rushing and roaring of the hail, as it approached, was truly tremendous.

Though the clouds at South Lambeth, near London, were at that juncture thin and light, and no storm was in sight, nor within hearing, yet the air was strongly electric; for the bells of an electric machine at that place rang repeatedly, and fierce sparks were discharged.

White's Monument

When I first took the present work in hand I proposed to have added an '*Annus Historico-naturalis*, or The Natural History of the Twelve Months of the Year'; which would have comprised many incidents and occurrences that have not fallen in my way to be mentioned in my series of letters; but, as Mr Aikin of Warrington has lately published somewhat of this sort, and as the length of my correspondence has sufficiently put your patience to the test, I shall here take a respectful leave of you and natural history together, and am,

With all due deference and regard,

Your most obliged and most humble servant,

GIL. WHITE

Selborne, June 25th, 1787

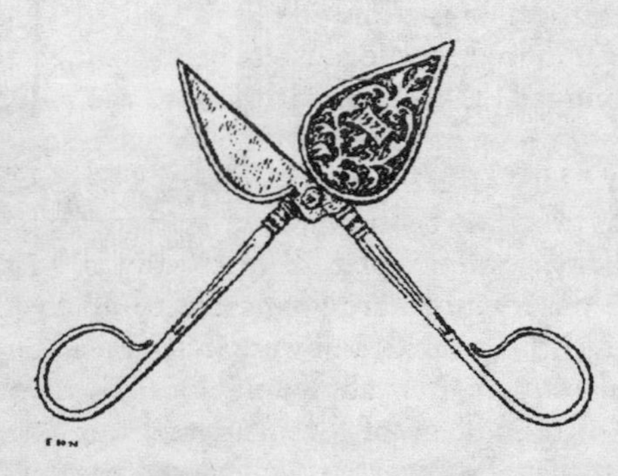

Advertisement

The advertisement to the octavo edition of *Selborne*, published in 1802, and edited by John White, the brother of the author, will explain the manner in which the following Calendar and Observations first came to be printed. I include them here in accordance with a now time-honoured custom. ED

The favourable reception with which the works on natural history of my late respected relation, the Revd Gilbert White of Selborne, have been honoured by the persons best qualified to judge of their merit, has induced me to present them to the public in a collected and commodious form, free from the encumbrance of any extraneous matter. His largest work, entitled *The Natural History of Selborne*, has probably been supposed by many to be formed upon a more local and confined plan than it really is. In fact, the greater part of the observations are applicable to all that portion of the island in which he resided, and were indeed made in various places. Almost the only matter absolutely local is the account of the antiquities of the village of Selborne; and this seemed to stand so much apart, that, however well calculated to gratify the lovers of topographical studies, it was thought that its entire omission would be considered no loss to the work, considered as a publication on natural history. Its place is occupied by the *Naturalists' Calendar, and Miscellaneous Observations*, which appeared in a separate volume since the author's decease, extracted from his papers by Dr Aikin. That gentleman has also made some farther selections from the papers, which are now all in my possession; and has undertaken the revision and arrangement of the whole. A very valuable addition to the calendar and observations has been obtained from the kindness of William Markwick, Esq., F.L.S., well known as an accurate observer of nature, whose parallel calendar, kept in the county of Sussex, is given upon the opposite columns.

The editor flatters himself that the publication in its present form will prove an acceptable addition to the library of the naturalist; and will in particular be useful in inspiring young persons, and those who pass their time in retirement, with a taste for the very pleasing branch of knowledge on which it treats.

J. W.

Fleet Street, 1802

OBSERVATIONS ON VARIOUS PARTS OF NATURE FROM MR WHITE'S MANUSCRIPT WITH REMARKS BY MR MARKWICK

The Wakes

Observations on Birds

Birds in General

In severe weather, fieldfares, redwings, skylarks, and tit-larks, resort to watered meadows for food; the latter wades up to its belly in pursuit of the pupae of insects, and runs along upon the floating grass and weeds. Many gnats are on the snow near the water; these support the birds in part.

Birds are much influenced in their choice of food by colour, for though white currants are a much sweeter fruit than red, yet they seldom touch the former till they have devoured every bunch of the latter.

Red-starts, fly-catchers, and black-caps, arrive early in April. If these little delicate beings are birds of passage (as we have reason to suppose they are, because they are never seen in winter), how could they, feeble as they seem, bear up against such storms of snow and rain, and make their way through such meteorous turbulences, as one should suppose would embarrass and retard the most hardy and resolute of the winged nation? Yet they keep their appointed times and seasons; and in spite of frosts and winds return to their stations periodically as if they had met with nothing to obstruct them. The withdrawing and appearance of the shortwinged summer birds is a very puzzling circumstance in natural history.

When the boys bring me wasps' nests, my bantam fowls fare deliciously, and when the combs are pulled to pieces, devour the young wasps in their maggot state with the highest glee and delight. Any insect-eating bird would do the same; and therefore I have often wondered that the accurate Mr Ray should call one species of buzzard *buteo apivorus sive vespivorus*, or the *honey buzzard*, because some combs of wasps happened to be found in one of their nests. The combs were conveyed thither doubtless for the sake of the maggots or nymphs, and not for their honey, since none is to be found in the combs of wasps. Birds of prey occasionally feed on insects; thus have I seen a tame kite picking up the female ants full of eggs, with much satisfaction. WHITE

That red-starts, fly-catchers, black-caps, and other slender-billed insectivorous small birds, particularly the swallow tribe, make their first appearance very early in the spring, is a well-known fact; though the fly-catcher is the latest of them all in its visit (as this accurate naturalist observes in another place), for it is never seen before the month of May. If these delicate creatures come to us from a distant country, they will probably be exposed in their passage, as Mr White justly remarks, to much greater difficulties from storms and tempests than their feeble powers appear to be able to surmount: on the other hand, if we suppose them to pass the winter in a dormant state in this country, concealed in caverns or other hiding-places sufficiently guarded from the extreme cold of our winter to preserve their life, and that at the approach of spring they revive from their torpid state and reassume their usual powers of action, it will entirely remove the first difficulty, arising from the storms and tempests they are liable to meet with in their passage; but how are we to get over the still greater difficulty of their revivification from their torpid state? What degree of warmth in the temperature of the air is necessary to produce that effect, and how it operates on the functions of animal life, are questions not easily answered.

How could Mr White suppose that Ray named this species the honey buzzard, because it fed on honey, when he not only named it in Latin *buteo apivorus et vespivorus*, but expressly says that 'it feeds on insects, and brings up its young with the maggots or nymphs of wasps'?

That birds of prey, when in want of their proper food, flesh, sometimes feed on insects I have little doubt, and I think I have observed the common buzzard, *falco buteo*, to settle on the ground and pick up insects of some kind or other. MARKWICK

Rooks

Rooks are continually fighting, and pulling each other's nests to pieces: these proceedings are inconsistent with living in such close community. And yet if a pair offer to build on a single tree, the nest is plundered and demolished at once. Some rooks roost on their nest trees. The twigs which the rooks drop in building supply the poor with brushwood to light their fires. Some unhappy pairs are not permitted to finish any nest till the rest have completed their building. As soon as they get a few sticks together, a party comes and demolishes the whole. As soon as rooks have finished their nests, and before they lay, the cocks begin to feed the hens, who receive their bounty with a fondling tremulous voice and fluttering wings, and all the little blandishments that are expressed by the young, while in a helpless state. This gallant deportment of the males is continued through the whole season of incubation. These birds do not copulate on trees, nor in their nests, but on the ground in the open fields. WHITE

After the first brood of rooks are sufficiently fledged, they all leave their nest-trees in the day-time, and resort to some distant place in search of food, but return regularly every evening, in vast flights, to their nest trees, where, after flying round several times with much noise and clamour till they are all assembled together, they take up their abode for the night. MARKWICK

Thrushes

Thrushes during long droughts are of great service in hunting out shell snails, which they pull to pieces for their young, and are thereby very serviceable in gardens. Missel thrushes do not destroy the fruit in gardens like the other species of *turdi*, but feed on the berries of mistletoe, and in the spring on ivy berries, which then begin to ripen. In the summer, when their young become fledged, they leave neighbourhoods, and retire to sheep-walks and wild commons.

The magpies, when they have young, destroy the broods of missel thrushes, though the dams are fierce birds, and fight boldly in defence of their nests. It is probably to avoid such insults, that this species of thrush, though wild at other times, delights to build near houses, and in frequented walks and gardens. WHITE

Of the truth of this I have been an eye-witness, having seen the common thrush feeding on the shell snail.

In the very early part of this spring (1797) a bird of this species used to sit every morning on the top of some high elms close by my windows, and delight me with its charming song, attracted thither probably, by some ripe ivy berries that grew near the place.

I have remarked something like the latter fact, for I remember, many years ago, seeing a pair of these birds fly up repeatedly and attack some larger bird, which I suppose disturbed their nest in my orchard, uttering at the same time violent shrieks. Since writing the above, I have seen more than once a pair of these birds attack some magpies that had disturbed their nest, with great violence and loud shrieks. WHITE

Poultry

Many creatures are endowed with a ready discernment to see what will turn to their own advantage and emolument: and often discover more sagacity than could be expected. Thus my neighbour's poultry watch for waggons loaded with wheat, and running after them, pick up a number of grains which are shaken from the sheaves by the agitation of the carriages. Thus, when my brother used to take down his gun to shoot sparrows, his cats would run out before him, to be ready to catch up the birds as they fell.

The earnest and early propensity of the *gallinae* to roost on high is very observable, and discovers a strong dread impressed on their spirits respecting vermin that may annoy them on the ground during the hours of darkness. Hence poultry, if left to themselves and not housed, will perch the winter through on yew-trees and fir-trees; and turkeys and guinea fowls, heavy as they are, get up into apple-trees; pheasants also in woods sleep on trees to avoid foxes; while pea-fowls climb to the tops of the highest trees round their owner's house for security, let the weather be ever so cold or blowing. Partridges, it is true, roost on the ground, not having the faculty of perching; but then the same fear prevails in their minds: for through apprehension from pole-cats and stoats, they never trust themselves to coverts, but nestle together in the midst of large fields, far removed from hedges and coppices, which they love to haunt in the day, and where at that season they can skulk more secure from the ravages of rapacious birds.

As to ducks and geese, their awkward splay web-feet forbid them to settle on trees: they therefore, in the hours of darkness and danger, betake themselves to their own element the water, where amidst large lakes and pools, like ships riding at anchor, they float the whole night long in peace and security. WHITE

Guinea fowls not only roost on high, but in hard weather resort, even in the daytime, to the very tops of the highest trees. Last winter, when the ground was covered with snow, I discovered all my guinea fowls, in the middle of the day, sitting on the highest boughs of some very tall elms, chattering and making a great clamour: I ordered them to be driven down lest they should be frozen to death in so elevated a situation, but this was not effected without much difficulty; they being very unwilling to quit their lofty abode, notwithstanding one of them had its feet so much frozen that we were obliged to kill it. I know not how to account for this, unless it was occasioned by their aversion to the snow on the ground, they being birds that come originally from a hot climate.

Notwithstanding the awkward splay web-feet (as Mr White calls them) of the duck genus, some of the foreign species have the power of settling on the boughs of trees apparently with great ease; an instance of which I have seen in the Earl of Ashburnham's menagerie, where the summer duck, *anas sponsa*, flew up and settled on the branch of an oak-tree in my presence: but whether any of them roost on trees in the night, we are not informed by any author that I am acquainted with. I suppose not, but that, like the rest of the genus, they sleep on the water, where the birds of this genus are not always perfectly secure, as will appear from the following circumstance which happened in this neighbourhood a few years since, as I was credibly informed. A female fox was found in the morning drowned in the same pond in which were several geese, and it was supposed that in the night the fox swam into the pond to devour the geese, but was attacked by the gander, which, being most powerful in his own element, buffeted the fox with its wings about the head till it was drowned. MARKWICK

Hen Partridge

A hen partridge came out of a ditch, and ran along shivering with her wings and crying out as if wounded and unable to get from us. While the dam acted this distress, the boy who attended me saw her brood, that was small and unable to fly, run for shelter into an old fox-earth under the bank. So wonderful a power is instinct. WHITE

It is not uncommon to see an old partridge feign itself wounded and run along on the ground fluttering and crying before either dog or man, to draw them away from its helpless unfledged young ones. I have seen it often, and once in particular I saw a remarkable instance of the old bird's solicitude to save its brood. As I was hunting a young pointer, the dog ran on a brood of very small partridges: the old bird cried, fluttered, and ran tumbling along just before the dog's nose till she had drawn him to a considerable distance, when she took wing, and flew still farther off, but not out of the field: on this the dog returned to me, near which place the young ones lay concealed in the grass, which the old bird no sooner perceived than she flew back again to us, settled just before the dog's nose again, and by rolling and tumbling about, drew off his attention from her young, and thus preserved her brood a second time. I have also seen, when a kite has been hovering over a covey of young partridges, the old birds fly up at the bird of prey, screaming and fighting with all their might to preserve their brood.

MARKWICK

A Hybrid Pheasant

Lord Stawell sent me from the great lodge in the Hold a curious bird for my inspection. It was found by the spaniels of one of his keepers in a coppice, and shot on the wing. The shape, air, and habit of the bird, and the scarlet ring round the eyes, agreed well with the appearance of a cock pheasant; but then the head and neck, and breast, and belly were of a glossy black: and though it weighed three pounds three ounces and a half,[1] the weight of a full grown cock pheasant, yet there were no signs of any spurs on the legs, as is usual with all grown cock pheasants, who have long ones. The legs and feet were naked of feathers and therefore it could be nothing of the grouse kind. In the tail were no bending feathers such as cock pheasants usually have, and are characteristic of the sex. The tail was much shorter than the tail of a hen pheasant, and blunt and square at the end. The back, wing feathers, and tail, were all of a pale russet, curiously streaked, somewhat like the upper parts of a hen partridge. I returned it with my verdict, that it was probably a spurious or hybrid hen bird, bred between a cock pheasant and some domestic fowl. When I came to talk with the keeper who brought it, he told me that some pea-hens had been known last summer to haunt the coppices and coverts where this mule was found.

Mr Elmer, of Farnham, the famous game painter, was employed to take an exact copy of this curious bird. WHITE

[N.B. It ought to be mentioned, that some good judges have imagined this bird to have been a stray grouse or blackcock; it is however to be observed, that Mr W. remarks, that its legs and feet were naked, whereas those of the grouse are feathered to the toes.]

1 Hen pheasants usually weigh only two pounds ten ounces.

Mr Latham observes that 'pea-hens, after they have done laying, sometimes assume the plumage of the male bird', and has given a figure of the male-feathered pea-hen now to be seen in the Leverian Museum; and M. Salerne remarks, that 'the hen pheasant, when she has done laying and Sitting, will get the plumage of the male'. May not this hybrid pheasant (as Mr White calls it) be a bird of this kind? that is, an old hen pheasant which has just begun to assume the plumage of the cock. MARKWICK

Land-rail

A man brought me a land-rail or daker-hen, a bird so rare in this district, that we seldom see more than one or two in a season, and those only in autumn. This is deemed a bird of passage by all the writers; yet from its formation, seems to be poorly qualified for migration; for its wings are short, and placed so forward, and out of the centre of gravity, that it flies in a very heavy and embarrassed manner, with its legs hanging down; and can hardly be sprung a second time, as it runs very fast, and seems to depend more on the swiftness of its feet than on its flying.

When we came to draw it, we found the entrails so soft and tender in appearance, they might have been dressed like the ropes of a woodcock. The craw or crop was small and lank, containing a mucus; the gizzard thick and strong, and filled with small shell snails, some whole, and many ground to pieces through the attrition which is occasioned by the muscular force and motion of that intestine. We saw no gravels among the food: perhaps the shell snails might perform the functions of gravels or pebbles, and might grind one another. Land-rails used to abound formerly, I remember, in the low wet bean-fields of Christian Malford in North Wilts, and in the meadows near Paradise Gardens at Oxford, where I have often heard them cry crex, crex. The bird mentioned above weighed seven and a half ounces, was fat and tender, and in flavour like the flesh of a woodcock. The liver was very large and delicate. WHITE

Land-rails are more plentiful with us than in the neighbourhood of Selborne. I have found four brace in an afternoon, and a friend of mine lately shot nine in two adjoining fields; but I never saw them in any other season than the autumn.

That it is a bird of passage there can be little doubt, though Mr White thinks it poorly qualified for migration, on account of the wings being short, and not placed in the exact centre of gravity; how that may be I cannot say, but I know that its heavy sluggish flight is not owing to its inability to fly faster, for I have seen it fly very swiftly, although in general its actions are sluggish. Its unwillingness to rise proceeds, I imagine, from its sluggish disposition, and its great timidity, for it will sometimes squat so close to the ground as to suffer itself to be taken up by the hand, rather than rise; and yet it will at times run very fast.

What Mr White remarks respecting the small shell snails found in its gizzard, confirms my opinion, that it frequents corn-fields, seed clover, and brakes or fern, more for the sake of snails, slugs, and other insects which abound in such places, than for the grain or seeds; and that it is entirely an insectivorous bird. WHITE

Food of the Ring-dove

One of my neighbours shot a ring-dove on an evening as it was returning from feed and going to roost. When his wife had picked and drawn it, she found its craw stuffed with the most nice and tender tops of turnips. These she washed and boiled, and so sat down to a choice and delicate plate of greens, culled and provided in this extraordinary manner.

Hence we may see that graminivorous birds, when grain fails can subsist on the leaves of vegetables. There is reason to suppose that they would not long be healthy without; for turkeys, though corn-fed, delight in a variety of plants, such as cabbage, lettuce, endive, &c., and poultry pick much grass; while geese live for months together on commons by grazing alone.

Nought is useless made;
On the barren heath
The shepherd tends his flock that daily crop
Their verdant dinner from the mossy turf
Sufficient: after them the cackling goose,
Close-grazier, finds wherewith to ease her want.

PHILIPS'S *Cyder*

WHITE

That many graminivorous birds feed also on the herbage or leaves of plants, there can be no doubt: partridges and larks frequently feed on the green leaves of turnips, which give a peculiar flavour to their flesh that is to me very palatable: the flavour also of wild ducks and geese greatly depends on the nature of their food;

and their flesh frequently contracts a rank unpleasant taste from their having lately fed on strong marshy aquatic plants, as I suppose.

That the leaves of vegetables are wholesome and conducive to the health of birds seems probable, for many people fat their ducks and turkeys with the leaves of lettuce chopped small. MARKWICK

Hen-harrier

A neighbouring gentleman sprung a pheasant in a wheat stubble, and shot at it; when, notwithstanding the report of the gun, it was immediately pursued by the blue hawk, known by the name of the hen-harrier, but escaped into some covert. He then sprung a second, and a third, in the same field, that got away in the same manner: the hawk hovering round him all the while that he was beating the field, conscious no doubt of the game that lurked in the stubble. Hence we may conclude that this bird of prey was rendered very daring and bold by hunger, and that hawks cannot always seize their game when they please. We may farther observe, that they cannot pounce their quarry on the ground where it might be able to make a stout resistance, since so large a fowl as a pheasant could not but be visible to the piercing eye of a hawk, when hovering over the field. Hence that propensity of cowering and squatting till they are almost trod on, which no doubt was intended as a mode of security; though long rendered destructive to the whole race of *gallinae* by the invention of nets and guns. WHITE

Of the great boldness and rapacity of birds of prey when urged on by hunger, I have seen several instances; particularly, when shooting in the winter in company with two friends, a woodcock flew across us, closely pursued by a small hawk: we all three fired at the woodcock instead of the hawk, which, notwithstanding the report of three guns close by it, continued its pursuit of the woodcock, struck it down, and carried it off, as we afterwards discovered.

At another time, when partridge-shooting with a friend, we saw a ring-tail hawk rise out of a pit with some large bird in its claws; though at a great distance; we both fired and obliged it to drop its prey, which proved to be one of the partridges which we were in pursuit of; and lastly, in an evening, I shot at and plainly saw that I had wounded a partridge, but it being late, was obliged to go home without finding it again. The next morning I walked round my land without any gun, but a favourite old spaniel followed my heels. When I came near the field where I wounded the bird the evening before, I heard the partridges call, and seeming to be much disturbed. On my approaching the bar-way, they all rose, some on my right, and some on my left hand; and just before and over my head, I perceived (though indistinctly from the extreme velocity of their motion) two birds fly directly against each other, when instantly, to my great astonishment, down dropped a partridge at my feet; the dog immediately seized it, and on examination, I found the blood flow very fast from a fresh wound in the head, but there was some dry clotted blood on its wings and side; whence I concluded that a hawk had singled out my wounded bird as the object of his prey, and had struck it down the instant that my approach had obliged the birds to rise on the wing; but the space between the hedges was so small, and the motion of the birds so instantaneous and quick, that I could not distinctly observe the operation. MARKWICK

Great Speckled Diver, or Loon

As one of my neighbours was traversing Wolmer forest from Bramshot across the moors, he found a large uncommon bird fluttering in the heath, but not wounded which he brought home alive. On examination it proved to be *Colymbus glacialis*, Linn., the great speckled diver or loon, which is most excellently described in Willughby's *Ornithology*.

Every part and proportion of this bird is so incomparably adapted to its mode of life, that in no instance do we see the wisdom of God in the creation to more advantage. The head is sharp and smaller than the part of the neck adjoining, in order that it may pierce the water; the wings are placed forward, and out of the centre of gravity, for a purpose which shall be noticed hereafter; the thighs quite at the podex, in order to facilitate diving; and the legs are flat, and as sharp backwards almost as the edge of a knife, that in striking they may easily cut the water; while the feet are palmated, and broad for swimming, yet so folded up when advanced forward to take a fresh stroke, as to be full as narrow as the shank. The two exterior toes of the feet are longest; the nails flat and broad, resembling the human, which give strength, and increase the power of swimming. The foot, when expanded, is not at right angles to the leg or body of the bird: but the exterior part inclining towards the head, forms an acute angle with the body, the intention being not to give motion in the line of the legs themselves, but by the combined impulse of both in an intermediate line, the line of the body.

Bramshott Church

Most people know, that have observed at all, that the swimming of birds is nothing more than a walking in the water, where one foot succeeds the other as on the land; yet no one, as far as I am aware, has remarked that diving fowls, while under water, impel and row themselves forward by a motion of their wings, as well as by the impulse of their feet: but such is really the case, as any person may easily be convinced, who will observe ducks when hunted by dogs in a clear pond. Nor do I know that any one has given a reason why the wings of diving fowls are placed so forward: doubtless, not for the purpose of promoting their speed in flying, since that position certainly impedes it; but probably for the increase of their motion under water, by the use of four oars instead of two; yet were the wings and feet nearer together, as in land-birds, they would, when in action, rather hinder than assist one another.

This colymbus was of considerable bulk, weighing only three drachms short of three pounds avoirdupois. It measured in length from the bill to the tail (which was very short) two feet, and to the extremities of the toes four inches more; and the breadth of the wings expanded was forty-two inches. A person attempted to eat the body, but found it very strong and rancid, as is the flesh of all birds living on fish. Divers or loons, though bred in the most northerly parts of Europe, yet are seen with us in very severe winters; and on the Thames they are called sprat loons, because they prey much on that sort of fish.

The legs of the *colymbi* and *mergi* are placed so very backward, and so out of all centre of gravity, that these birds cannot walk at all. They are called by Linnaeus *compedes*, because they move on the ground as if shackled or fettered. WHITE

These accurate and ingenious observations, tending to set forth in a proper light the wonderful works of God in the creation, and to point out His wisdom in adapting the singular form and position of the limb of this bird to the particular mode in which it is destined to pass the greatest part of its life in an element much denser than the air, do Mr White credit, not only as a naturalist, but as a man and as a philosopher, in the truest sense of the word, in my opinion; for were we enabled to trace the works of nature minutely and accurately, we should find, not only that every bird, but every creature, was equally well adapted to the purpose for which it was intended; though this fitness and propriety of form is more striking in such animals as are destined to any uncommon mode of life.

I have had in my possession two birds, which, though of a

different genus, bear a great resemblance to Mr White's *colymbus*, in their manner of life, which is spent chiefly in the water, where they swim and dive with astonishing rapidity, for which purpose their fin-toed feet, placed far behind, and very short wings, are particularly well adapted, and show the wisdom of God in the creation as conspicuously as the bird before mentioned. These birds were the greater and lesser crested grebe, *podiceps cristatus et auritus*. What surprised me most was, that the first of these birds was found alive on dry ground, about seven miles from the sea, to which place there was no communication by water. How did it get so far from the sea? its wings and legs being so ill adapted either to flying or walking. The lesser crested grebe was also found in a fresh water pond which had no communication with other water at some miles' distance from the sea. MARKWICK

Stone-curlew

On the 27th of February, 1788, stone-curlews were heard to pipe: and on March 1st, after it was dark, some were passing over the village, as might be perceived by their quick short note, which they use in their nocturnal excursions by way of watchword, that they may not stray and lose their companions.

Thus, we see, that retire whithersoever they may in the winter, they return again early in the spring, and are, as it now appears, the first summer birds that come back. Perhaps the mildness of the season may have quickened the emigration of the curlews this year.

They spend the day in high elevated fields and sheep-walks; but seem to descend in the night to streams and meadows, perhaps for water, which their upland haunts do not afford them. WHITE

On the 31st of January, 1792, I received a bird of this species which had been recently killed by a neighbouring farmer, who said he had frequently seen it in his fields during the former part of the winter: this perhaps was an occasional straggler, which by some accident was prevented from accompanying its companions in their migration.
MARKWICK

The smallest uncrested Willow Wren

The smallest uncrested willow wren, or chiff-chaff, is the next early summer bird which we have remarked; it utters two sharp piercing notes, so loud in hollow woods, as to occasion an echo, and is usually first heard about the 20th of March. WHITE

This bird, which Mr White calls the smallest willow wren or chiff-chaff, makes its appearance very early in spring, and is very common with us, but I cannot make out the three different species of willow wrens which he assures us he has discovered. Ever since the publication of his History of Selborne I have used my utmost endeavours to discover his three birds, but hitherto without success. I have frequently shot the bird which 'haunts only the tops of trees, and makes a sibilous noise', even in the very act of uttering that sibilous note, but it always proved to be the common willow wren or his chiff-chaff. In short, I never could discover more than one species, unless my greater petty-chaps, *Sylvia hortensis* of Latham, is his greatest willow wren. MARKWICK

Fern-owl or Goat Sucker

The country people have a notion that the fern-owl, or churn-owl, or eve-jarr, which they also call a puckeridge, is very injurious to weaning calves, by inflicting as it strikes at them, the fatal distemper known to cow-leeches by the name of puckeridge. Thus does this harmless ill-fated bird fall under a double imputation which it by no means deserves – in Italy, of sucking the teats of goats, whence it is called *caprimulgus*; and with us, of communicating a deadly disorder to cattle. But the truth of the matter is, the malady above-mentioned is occasioned by the *Aestrus bovis*, a dipterous insect, which lays its eggs along the chines of kine, where the maggots, when hatched, eat their way through the hide of the beast into the flesh, and grow to a very large size. I have just talked with a man who says he has more than once stripped calves who have died of the puckeridge; that the ail or complaint lay along the chine, where the flesh was much swelled, and filled with purulent matter. Once I myself saw a large rough maggot of this sort squeezed out of the back of a cow.

These maggots in Essex are called wornils.

The least observation and attention would convince men that these birds neither injure the goatherd nor the grazier, but are perfectly harmless, and subsist alone, being night birds, on night insects, such as *Scarabaei* and *Phalaenae*; and through the month of July mostly on the *Scarabaeus solstitialis*, which in many districts abounds at that season. Those that we have opened, have always had their craws stuffed with large night moths and their eggs, and pieces of chaffers: nor does it anywise appear how they can, weak and unarmed as they seem, inflict any harm upon kine, unless they possess the powers of animal magnetism and can affect them by fluttering over them.

A fern-owl this evening (August 27) showed off in a very unusual and entertaining manner, by hawking round and round the circumference of my great spreading oak for twenty times following, keeping mostly close to the grass, but occasionally glancing up amidst the boughs of the tree. This amusing bird was then in pursuit of a brood of some particular *phalaena* belonging to the oak, of which there are several sorts; and exhibited on the occasion a command of wing superior, I think, to that of the swallow itself.

When a person approaches the haunt of fern-owls in an evening, they continue flying round the head of the obtruder; and by striking their wings together above their backs, in the manner that the pigeons called smiters are known to do, made a smart snap; perhaps at that time they are jealous for their young, and their noise and gesture are intended by way of menace.

Fern-owls have attachment to oaks, no doubt on account of food; for the next evening we saw one again several times among the boughs of the same tree; but it did not skim round its stem over the grass, as on the evening before. In May these birds find the *Scarabaeus melolontha* on the oak, and the *Scarabaeus solstitialis* at mid-summer. These peculiar birds can only be watched and observed for two hours in the twenty-four; and then in dubious twilight an hour after sunset and an hour before sunrise.

On this day (July 14 1789) a woman brought me two eggs of a fern-owl or evening-jarr, which she found on the verge of the Hanger, to the left of the hermitage, under a beechen shrub. This person, who lives just at the foot of the Hanger, seems well acquainted with these nocturnal swallows, and says she has often found their eggs near that place, and that they lay only two at a time on the bare ground. The eggs were oblong, dusky, and streaked somewhat in the manner of the plumage of the parent bird and were equal in size at each end. The dam was sitting on the eggs when found, which contained the rudiments of young, and would have been hatched perhaps in a week. From hence we may see the time of their breeding, which corresponds pretty well with that of the swift, as does also the period of their arrival. Each species is usually seen about the beginning of May. Each breeds but once in a summer; each lays only two eggs.

July 4, 1790. The woman who brought me two fern-owl's eggs last year on July 14, on this day produced me two more, one of which had been laid this morning, as appears plainly, because there was only one in the nest the evening before. They were found, as last July, on the verge of the down above the hermitage under a beechen shrub, on the naked ground. Last year those eggs were full of young, just ready to be hatched.

These circumstances point out the exact time when these curious nocturnal migratory birds lay their eggs and hatch their young. Fern-owls, like snipes, stone-curlews, and some other birds, make no nest. Birds that build on the ground do not make much of nests.

WHITE

No author that I am acquainted with has given so accurate and pleasing an account of the manners and habits of the goat-sucker as Mr White, taken entirely from his own observations. Its being a nocturnal bird has prevented my having many opportunities of observing it. I suspect that it passes the day in concealment amidst the dark and shady gloom of deep-wooded dells, or as they are called here gills; having more than once seen it roused from such solitary places by my dogs, when shooting in the daytime. I have also sometimes seen it in an evening, but not long enough to take notice of its habits and manners. I have never seen it but in the summer, between the months of May and September. MARKWICK

Sand-martins

March 23, 1788. A gentleman, who was this week on a visit at Waverley, took the opportunity of examining some of the holes in the sandbanks with which that district abounds. As these are undoubtedly bored by bank-martins, and are the places where they avowedly breed, he was in hopes they might have slept there also, and that he might have surprised them just as they were awaking from their winter slumbers. When he had dug for some time he found the holes were horizontal and serpentine, as I had observed before; and that the nests were deposited at the inner end, and had been occupied by broods in former summers, but no torpid birds were to be found. He opened and examined about a dozen holes. Another gentleman made the same search many years ago, with little success.

These holes were in depth about two feet.

March 21, 1790. A single bank or sand-martin was seen hovering and playing round the sand-pit at Short Heath, where in the summer they abound.

April 9, 1793. A sober hind assures us that this day, on Wishhanger

Common, between Hedleigh and Frinsham, he saw several blank-martins playing in and out, and hanging before some nest-holes in a sand-hill, where these birds usually nestle.

The incident confirms my suspicions, that this species of hirundo is to be seen first of any; and gives great reason to suppose that they do not leave their wild haunts at all, but are secreted amidst the clefts and caverns of those abrupt cliffs, where they usually spend their summers.

The late severe weather considered, it is not very probable that these birds should have migrated so early from a tropical region, through all these cutting winds and pinching frosts; but it is easy to suppose that they may, like bats and flies, have been awakened by the influence of the sun, amidst their secret latebrae, where they have spent the uncomfortable foodless months in a torpid state and the profoundest of slumbers.

There is a large pond at Wishhanger, which induces these sandmartins to frequent that district. For I have ever remarked that they haunt near great waters, either rivers or lakes. WHITE

Here, and in many other passages of his writings, this very ingenious naturalist savours the opinion that part at least of the swallow tribe pass their winter in a torpid state in the same manner as bats and flies, and revive again on the approach of spring.

I have frequently taken notice of all these circumstances, which induces Mr White to suppose that some of these hirundines lie torpid during winter. I have seen so late as November, on a finer day than usual at that season of the year, two or three swallows flying backwards and forwards under a warm hedge, or on the sunny side of some old building: nay, I once saw on the 8th of December two martins flying about very briskly, the weather being mild. I had not seen any considerable number either of swallows or martins for a considerable time before; from whence then, could these few birds come, if not from some hole or cavern where they had laid themselves up for the winter? Surely it will not be asserted that these birds migrate back again from some distant tropical region merely on the appearance of a fine day or two at this late season of the year. Again, very early in the spring, and sometimes immediately after very cold severe weather, on its growing a little warmer, a few of these birds suddenly make their appearance, long before the generality of them are seen. These appearances certainly favour the opinion of their passing the winter in a torpid state, but do not absolutely prove the fact; for who ever saw them reviving of their own accord from their torpid state, without being first brought to the fire, and as it were forced into life again? soon after which revivification they constantly die. MARKWICK

Swallows, congregating and disappearance of

During the severe winds that often prevail late in the spring it is not easy to say how the hirundines subsist; for they withdraw themselves, and are hardly ever seen, nor do any insects appear for their support. That they can retire to rest and sleep away these uncomfortable periods, as bats do, is a matter rather to be suspected than proved; or do they not rather spend their time in deep and sheltered vales near waters, where insects are more likely to be found? Certain it is, that hardly any individuals of this genus have at such times been seen for several days together.

September 13, 1791. The congregating flocks of hirundines on the church and tower are very beautiful and amusing. When they fly off together from the roof, on any alarm, they quite swarm in the air. But they soon settle in heaps, and preening their feathers, and lifting up their wings to admit the sun, seem highly to enjoy the warm situation. Thus they spend the heat of the day preparing for their emigration, and, as it were, consulting when and where they are to go. The flight about the church seems to consist chiefly of house-martins, about 400 in number; but there are other places of rendezvous about the village frequented at the same time.

It is remarkable that though most of them sit on the battlements and roof, yet many hang or cling for some time by their claws against the surface of the walls, in a manner not practised by them at any other time of their remaining with us.

The swallows seem to delight more in holding their assemblies on trees.

November 3, 1789. Two swallows were seen this morning at Newton vicarage-house, hovering and settling on the roofs and outbuildings. None have been observed at Selborne since October 11. It is very remarkable, that after the hirundines have disappeared for some weeks, a few are occasionally seen again; sometimes in the first week in November, and that only for one day. Do they not withdraw and slumber in some hiding-place in the interval? For we cannot suppose they had emigrated to warmer climes and so returned again for one day. Is it not more probable that they are awakened from sleep, and, like the bats, are come forth to collect a little food? Bats appear at all seasons through the autumn and spring months, when the thermometer is at 50, because then *phalaenae* and moths are stirring.

These swallows looked like young ones. WHITE

Of their migration the proofs are such as will scarcely admit of a doubt. Sir Charles Wager and Captain Wright saw vast flocks of them at sea, when on their passage from one country to another. Our author, Mr White, saw what he deemed the actual migration of these birds, and which he has described at p. 259 of his *History of Selborne*; and of their congregating together on the roofs of churches and other buildings, and on trees, previous to their departure, many instances occur; particularly I once observed a large stock of house-martins on the roof of the church here at Catsfield, which acted exactly in the manner here described by Mr White, sometimes preening their feathers and spreading their wings to the sun, and then flying off all together, but soon returning to their former situation. The greatest part of these birds seem to be young ones.

MARKWICK

Wagtails

While the cows are feeding in the moist low pastures, broods of wagtails, white and grey, run round them, close up to their noses, and under their very bellies, availing themselves of the flies that settle on their legs, and probably finding worms and *larvae* that are roused by the trampling of their feet. Nature is such an economist, that the most incongruous animals can avail themselves of each other.

Interest makes strange friendships. WHITE

Birds continually avail themselves of particular and unusual circumstances to procure their food; thus wagtails keep playing about the noses and legs of cattle as they feed, in quest of flies and other insects which abound near those animals; and great numbers of them will follow close to the plough to devour the worms, &c., that are turned up by that instrument. The redbreast attends the gardener when digging his borders; and will, with great familiarity and tameness, pick out the worms, almost close to his spade, as I have frequently seen. Starlings and magpies very often sit on the backs of sheep and deer to pick out their ticks. MARKWICK

Wryneck

These birds appear on the grass-plots and walks; they walk a little as well as hop, and thrust their bills into the turf, in quest, I conclude, of ants, which are their food. While they hold their bills in the grass they draw out their prey with their tongues, which are so long as to be coiled round their heads. WHITE

Grosbeak

Mr B. shot a cock grosbeak which he had observed to haunt his garden for more than a fortnight. I began to accuse this bird of making sad havoc among the buds of the cherries, gooseberries, and wall-fruit of all the neighbouring orchards. Upon opening its crop or craw no buds were to be seen, but a mass of kernels of the stones of fruits. Mr B. observed that this bird frequented the spot where plum-trees grow, and that he had seen it with somewhat hard in its mouth, which it broke with difficulty; these were the stones of damsons. The Latin ornithologists call this bird *Coccothraustes* – *i.e.*, berry-breaker, because with its large horny beak it cracks and breaks the shells of stone-fruits for the sake of the seed or kernel. Birds of this sort are rarely seen in England, and only in winter.

WHITE

I have never seen this rare bird but during the severest cold of the hardest winters; at which season of the year I have had in my possession two or three that were killed in this neighbourhood in different years. MARKWICK

Observations on Quadrupeds

Sheep

The sheep on the downs this winter (1769) are very ragged, and their coats much torn; the shepherds say they tear their fleeces with their own mouths and horns, and they are always in that way in mild wet winters, being teased and tickled with a kind of lice.

After ewes and lambs are shorn, there is great confusion and bleating, neither the dams nor the young being able to distinguish one another as before. This embarrassment seems not so much to arise from the loss of the fleece, which may occasion an alteration in their appearance, as from the defect of that *notus odor*, discriminating each individual personally; which also is confounded by the strong scent of pitch and tar wherewith they are newly marked; for the brute creation recognise each other more from the smell than the sight; and in matters of identity and diversity appeal much more to their noses than their eyes. After sheep have been washed there is the same confusion, from the reason given above. WHITE

Rabbits

Rabbits make incomparably the finest turf, for they not only bite closer than larger quadrupeds, but they allow no bents to rise; hence warrens produce much the most delicate turf for gardens. Sheep never touch the stalks of grasses. WHITE

Cat and Squirrels

A boy has taken three young squirrels in their nest or drey as it is called in these parts. These small creatures he put under the care of a cat who had lately lost her kittens, and finds that she nurses and suckles them with the same assiduity and affection as if they were her own offspring. This circumstance corroborates my suspicion that the mention of exposed and deserted children being nurtured by female beasts of prey who had lost their young may not be so improbable an incident as many have supposed; and therefore may be a justification of those authors who have gravely mentioned what some have deemed to be a wild and improbable story.

So many people went to see the little squirrels suckled by a cat that the foster-mother became jealous of her charge, and in pain for their safety; and therefore hid them over the ceiling, where one died. This circumstance shows her affection for these fondlings, and that she supposes the squirrels to be her own young. Thus hens, when they have hatched ducklings, are equally attached to them as if they were their own chickens. WHITE

Horse

An old hunting mare, which ran on the common, being taken very ill, ran down into the village, as it were, to implore the help of men and died the night following in the street. WHITE

Hounds

The king's stag-hounds came down to Alton, attended by a huntsman and six yeomen prickers, with horns, to try for the stag that has haunted Hartley Wood for so long a time. Many hundreds of people, horse and foot, attended the dogs to see the deer unharboured; but though the huntsmen drew Hartley Wood and Long Coppice, and Shrubwood, and Temple Hangers, and in their way back Hartley and Wardleham Hangers, yet no stag could be found.

The royal pack, accustomed to have the deer turned out before them, never drew the coverts with any address and spirit, as many people that were present observed; and this remark the event has proved to be a true one. For as a person was lately pursuing a pheasant that was wing-broken in Hartley Wood, he stumbled upon the stag by accident, and ran in upon him as he lay concealed amidst a thick brake of brambles and bushes. WHITE

Observations on Insects and Vermes

Insects in general

The day and night insects occupy the annuals alternately: the papilios, muscae, and apes, are succeeded at the close of day by phalaenae, earwigs, woodlice, &c. In the dusk of the evening, when beetles begin to buzz, partridges begin to call; these two circumstances are exactly coincident.

Ivy is the last flower that supports the hymenopterous and dipterous insects. On sunny days quite on to November, they swarm on trees covered with this plant; and when they disappear, probably retire under the shelter of its leaves, concealing themselves between its fibres and the trees which it entwines. WHITE

This I have often observed, having seen bees and other winged insects swarming about the flowers of the ivy very late in the autumn. MARKWICK

Spiders, woodlice, lepismae in cupboards and among sugar, some empedes, gnats, flies of several species, some phalaenae in hedges, earth worms, &c., are stirring at all times when winters are mild, and are of great service to those soft-billed birds that never leave us.

On every sunny day the winter through clouds of insects usually called gnats (I suppose tipulae and empedes) appear sporting and dancing over the tops of the evergreen-trees in the shrubbery, and striking about as if the business of generation was still going on. Hence it appears that these diptera (which by their sizes appear to be of different species), are not subject to a torpid state in the winter, as most winged insects are. At night, and in frosty weather, and when it rains and blows, they seem to retire into those trees. They often are out in a fog. WHITE

This I have also seen, and have frequently observed swarms of little winged insects playing up and down in the air in the middle of winter, even when the ground has been covered with snow.

MARKWICK

Humming in the air

There is a natural occurrence to be met with upon the highest part of our own down in hot summer days, which always amuses me much, without giving me any satisfaction with respect to the cause of it; and that is, a loud audible humming of bees in the air, though not one insect is to be seen. This sound is to be heard distinctly the whole common through, from the Money-dells to Mr White's avenue gate. Any person would suppose that a large swarm of bees was in motion, and playing about over his head. This noise was heard last week, on June 28th.

> Resounds the living surface of the ground,
> Nor undelightful is the ceaseless hum
> To him who muses – at noon . . .
> Thick in yon stream of light a thousand ways,
> Upward and downward, thwarting and convolv'd,
> The quivering nations sport. – THOMPSON's *Seasons*

WHITE

Chaffers

Cockchaffers seldom abound oftener than once in three or four years; when they swarm, they deface the trees and hedges. Whole woods of oaks are stripped bare by them.

Chaffers are eaten by the turkey, the rook, and the house-sparrow.

The *Scarabaeus solstitialis* first appears about June 26th: they are very punctual in their coming out every year. They are a small species, about half the size of the Maychaffer, and are known in some parts by the name of the fern chaffer. WHITE

A singular circumstance relative to the cockchaffer, or as it is called here, the May-bug (*Scarabaeus melolontha*), happened this year (1800): My gardener, in digging some ground, found, about six inches under the surface, two of these insects alive and perfectly formed, so early as the 24th of March. When he brought them to me, they appeared to be as perfect and as much alive as in the midst of summer, crawling about as briskly as ever: yet I saw no more of this insect till the 22nd of May, when it began to make its appearance. How comes it, that though it was perfectly formed so early as the 24th March, it did not show itself above ground till nearly two months afterwards? MARKWICK

Ptinus Pectinicornis

Those maggots that make worm-holes in tables, chairs, bedposts, &c., and destroy wooden furniture, especially where there is any sap, are the larvae of the *Ptinus pectinicornis.* This insect, it is probable, deposits its eggs on the surface, and the worms eat their way in.

In their holes they turn into their pupae state, and so come forth winged in July; eating their way through the valances or curtains of a bed, or any other furniture that happens to obstruct their passage.

They seem to be most inclined to breed in beech: hence beech will not make lasting utensils or furniture. If their eggs are deposited on the surface, frequent rubbing will preserve wooden furniture.

WHITE

Blatta Orientalis – Cockroach

A neighbour complained that her house was overrun with a kind of blackbeetle, or, as she expressed herself, with a kind of blackbob, which swarmed in her kitchen when they got up in a morning before daybreak.

Soon after this account I observed an unusual insect in one of my dark chimney closets, and find since, that in the night they swarm also in my kitchen. On examination I soon ascertained the species to be the *Blatta orientalis* of Linnaeus, and the *Blatta molendinaria* of Mouffet. The male is winged; the female is not, but shows somewhat like the rudiments of wings, as if in the pupa state.

These insects belonged originally to the warmer parts of America, and were conveyed from thence by shipping to the East Indies; and by means of commerce begin to prevail in the more northern parts of Europe, as Russia, Sweden, &c. How long they have abounded in England I cannot say; but have never observed them in my house till lately.

They love warmth, and haunt chimney closets and the backs of ovens. Poda says that these and house crickets will not associate together; but he is mistaken in that assertion, as Linnaeus suspected he was. They are altogether night insects, (*Lucifugae*), never coming forth till the rooms are dark and still, and escaping away nimbly at the approach of a candle. Their antennae are remarkably long, slender, and flexible.

October, 1790. After the servants are gone to bed the kitchen, hearth swarms with the young crickets and young *Blattae molendinariae* of all sizes, from the most minute growth to their full proportions. They seem to live in a friendly manner together, and not to prey the one on the other.

August, 1792. After the destruction of many thousands of *Blattae molendinariae*, we find that at intervals a fresh detachment of old ones arrives, and particularly during this hot season; for the windows being left open in the evenings, the males come flying in at the casements from the neighbouring houses, which swarm with them. How the females, that seem to have no perfect wings that they can use, can contrive to get from house to house does not so readily appear. These, like many insects, when they find their present abodes overstocked, have powers of migrating to fresh quarters. Since the *Blattae* have been so much kept under, the crickets have greatly increased in number. WHITE

Gryllus Domesticus – House Cricket

November. After the servants are gone to bed the kitchen hearth swarms with minute crickets not so large as fleas, which must have been lately hatched. So that these domestic insects, cherished by the influence of a constant large fire, regard not the season of the year, but produce their young at a time when their congeners are either dead or laid up for the winter, to pass away the uncomfortable months in the profoundest slumbers, and a state of torpidity.

When house-crickets are out and running about in a room in the night, if surprised by a candle, they give two or three shrill notes, as it were for a signal to their fellows, that they may escape to their crannies and lurking-holes, to avoid danger. WHITE

Cimex Linearis

August 12, 1775. *Cimices lineares* are now in high copulation on ponds and pools. The females, who vastly exceed the males in bulk, dart and shoot along on the surface of the water with the males on their backs. When a female chooses to be disengaged, she rears, and jumps, and plunges, like an unruly colt; the lover thus dismounted, soon finds a new mate. The females, as fast as their curiosities are satisfied, retire to another part of the lake, perhaps to deposit their foetus in quiet; hence the sexes are found separate, except where generation is going on. From the multitude of minute young of all gradations of sizes, these insects seem without doubt to be viviparous.

WHITE

Phalaena Quercus

Most of our oaks are naked of leaves, and even the Holt in general, having been ravaged by the caterpillars of a small *Phalaena*, which is of a pale yellow colour. These insects, though a feeble race, yet, from their infinite numbers, are of wonderful effect, being able to destroy the foliage of whole forests and districts. At this season they leave their *aurelia*, and issue forth in their fly-state, swarming and covering the trees and hedges.

In a field at Greatham I saw a flight of swifts busied in catching their prey near the ground, and found they were hawking after these *Phalaenae*. The *aureliae* of this moth is shining and as black as jet, and lies wrapped up in a leaf of the tree, which is rolled round it, and secured at the ends by a web, to prevent the maggot from falling out. WHITE

I suspect that the insect here meant is not the *Phalaena quercus*, but the *Phalaena viridata*, concerning which I find the following note in my *Naturalist's Calendar* for the year 1785.

About this time, and for a few days last past, I observed the leaves of almost all the oak-trees in Denn copse to be eaten and destroyed, and, on examining more narrowly, saw an infinite number of small beautiful pale green moths flying about the trees; the leaves of which that were not quite destroyed were curled up, and withinside were the exuviae or remains of the *chrysalis*, from whence I suppose the moths had issued, and whose caterpillar had eaten the leaves. MARKWICK

Ephemera Cauda Biseta – May-fly

June 10, 1771. Myriads of May-flies appear for the first time on the Alresford stream. The air was crowded with them and the surface of the water covered. Large trouts sucked them in as they lay struggling on the surface of the stream, unable to rise till their wings were dried.

This appearance reconciled me in some measure to the wonderful account that Scopoli gives of the quantities emerging from the rivers of Carniola. Their motions are very peculiar, up and down for many yards almost in a perpendicular line. WHITE

I once saw a swarm of these insects playing up and down over the surface of a pond in Denn Park, exactly in the manner described by this accurate naturalist. It was late in the evening of a warm summer's day when I observed them. MARKWICK

Sphynx Ocellata

A vast insect appears after it is dusk, flying with a humming noise, and inserting its tongue into the bloom of the honeysuckle; it scarcely settles upon the plants, but feeds on the wing in the manner of humming-birds. WHITE

I have frequently seen the large bee moth, *Sphinx stellatarum*, inserting its long tongue or proboscis into the centre of flowers, and feeding on their nectar, without settling on them, but keeping constantly on the wing. MARKWICK

Wild Bee

There is a sort of wild bee frequenting the garden campion for the sake of its tomentum, which probably it turns to some purpose in the business of nidification. It is very pleasant to see with what address it strips off the pubes, running from the top to the bottom of a branch, and shaving it bare with all the dexterity of a hoop-shaver. When it has got a vast bundle, almost as large as itself, it flies away, holding it secure between its chin and its fore legs.

There is a remarkable hill on the downs near Lewes in Sussex, known by the name of Mount Carburn, which overlooks that town, and affords a most engaging prospect of all the country round, besides several views of the sea. On the very summit of this exalted promontory, and amidst the trenches of its Danish camp, there haunts a species of wild bee, making its nest in the chalky soil. When people approach the place, these insects begin to be alarmed, and, with a sharp and hostile sound, dash and strike round the heads and faces of intruders. I have often been interrupted myself while contemplating the grandeur of the scenery around me, and have thought myself in danger of being stung. WHITE

Wasps

Wasps abound in woody wild districts far from neighbourhoods; they feed on flowers, and catch flies and caterpillars to carry to their young. Wasps make their nests with the raspings of sound timber; hornets with what they gnaw from decayed: these particles of wood are kneaded up with a mixture of saliva from their bodies and moulded into combs.

When there is no fruit in the gardens, wasps eat flies, and suck the honey from flowers, from ivy blossoms and umbellated plants: they carry off also flesh from butchers' shambles. WHITE

In the year 1775, wasps abounded so prodigiously in this neighbourhood, that in the month of August no less than seven or eight of their nests were ploughed up in one field: of which there were several instances, as I was informed.

In the spring, about the beginning of April, a single wasp is sometimes seen, which is of a larger size than usual; this I imagine is the queen or female wasp, and the mother of the future swarm.

MARKWICK

Oestrus Curvicauda

This insect lays its nits or eggs on horses' legs, flanks, &c., each on a single hair. The maggots, when hatched, do not enter the horses' skins, but fall to the ground. It seems to abound most in moist, moorish places, though sometimes seen in the uplands. WHITE

Nose-fly

About the beginning of July, a species of fly (*musca*) obtains, which proves very tormenting to horses, trying still to enter their nostrils and ears, and actually laying their eggs in the latter of those organs, or perhaps in both. When these abound, horses in woodland districts, become very impatient at their work, continually tossing their heads, and rubbing their noses on each other, regardless of the driver, so that accidents often ensue. In the heat of the day, men are often obliged to desist from ploughing. Saddle-horses are also very troublesome at such seasons. Country people call this insect the nose-fly. WHITE

Is not this insect the *Oestrus nasalis* of Linnaeus, so well described by Mr Clark in the third volume of the *Linnaean Transactions*, under the name of *Oestrus veterinus?*. MARKWICK

Icheumon Fly

I saw lately a small ichneumon-fly attack a spider much larger than itself on a grass walk. When the spider made any resistance, the ichneumon applied her tail to him, and stung him with great vehemence, so that he soon became dead and motionless. The ichneumon then running backward drew her prey very nimbly over the walk into the standing grass. This spider would be deposited in some hole where the ichneumon would lay some eggs; and as soon as the eggs were hatched, the carcase would afford ready food for the maggots.

Perhaps some eggs might be injected into the body of the spider, in the act of stinging. Some ichneumon deposit their eggs in the aurelia of moths and butterflies. WHITE

In my *Naturalist's Calendar* for 1795, July 21st, I find the following note:

It is not uncommon for some of the species of ichneumon-flies to deposit their eggs in the chrysalis of a butterfly; some time ago I put two of the chrysales of a butterfly into a box, and covered it with gauze, to discover what species of butterfly they would produce; but instead of a butterfly, one of them produced a number of small ichneumon-flies.

There are many instances of the great service these little insects are to mankind in reducing the number of noxious insects, by depositing their eggs in the soft bodies of their *larvae*; but none more remarkable than that of the ichneumon *tipulae*, which pierces the tender bodies and deposits its eggs in the *larva* of the *Tipula tritici*, an insect, which, when it abounds greatly, is very prejudicial to the grains of wheat. This operation I have frequently seen it perform with wonder and delight. MARKWICK

Bombylius Medius

The *Bombylius medius is* much about in March and the beginning of April, and soon seems to retire. It is an hairy insect, like a humble-bee, but with only two wings, and a long straight beak, with which it sucks the early flowers. The female seems to lay its eggs as it poises on its wings, by striking its tail on the ground, and against the grass that stands in its way, in a quick manner, for several times together. WHITE

I have often seen this insect fly with great velocity, stop on a sudden hang in the air in a stationary position for some time, and then fly off again; but do not recollect having ever seen it strike its tail against the ground, or any other substance. MARKWICK

Muscae – Flies

In the decline of the year, when the mornings and evenings become chilly, many species of flies (*Muscae*) retire into houses, and swarm in the windows.

At first they are very brisk and alert; but as they grow more torpid, one cannot help observing that they move with difficulty, and are scarce able to lift their legs, which seem as if glued to the glass; and by degrees many do actually stick on till they die in the place.

It has been observed that divers flies, beside their sharp hooked nails, have also skinny palms, or flaps to their feet, whereby they are enabled to stick on the glass and other smooth bodies, and to walk on ceilings with their backs downward, by means of the pressure of the atmosphere on those flaps; the weight of which they easily overcome in warm weather, when they are brisk and alert. But in the decline of the year, this resistance becomes too mighty for their diminished strength; and we see flies labouring along, and lugging their feet in windows as if they stuck to the glass, and it is with the utmost difficulty they can draw one foot after another, and disengage their hollow caps from the slippery surface.

Upon the same principle that flies stick and support themselves; do boys, by way of play, carry heavy weights by only a piece of wet leather at the end of a string clapped close on the surface of a stone.

WHITE

Tipulae, or *Empedes*

May. Millions of *empedes*, or *tipulae*, come forth at the close of day, and swarm to such a degree as to fill the air. At this juncture they sport and copulate; as it grows more dark they retire. All day they hide in the hedges. As they rise in a cloud they appear like smoke.

I do not remember to have seen such swarms, except in the fens of the Isle of Ely. They appear most over grass grounds. WHITE

Aphides

On the 1st of August, about half an hour after three in the afternoon, the people of Selborne were surprised by a shower of aphides which fell in these parts. They who were walking in the streets at that time found themselves covered with these insects, which settled also on the trees and gardens, and blackened all the vegetables where they alighted. These armies, no doubt, were in a state of emigration, and shifting their quarters; and might perhaps come from the great hop-plantations of Kent or Sussex, the wind being that day at north. They were observed at the same time at Farnham, and all along the vale of Alton. WHITE

Ants

August 23. Every ant-hill about this time is in a strange hurry and confusion; and all the winged ants, agitated by some violent impulse, are leaving their homes, and, bent on emigration, swarm by myriads in the air, to the great emolument of the hirundines, which fare luxuriously. Those that escape the swallows return no more to their nests, but looking out for fresh settlements, lay a foundation for future colonies. All the females at this time are pregnant: the males that escape being eaten, wander away and die.

October 2. Flying-ants, male and female, usually swarm and migrate on hot sunny days in August and September; but this day a vast emigration took place in my garden, and myriads came forth, in appearance from the drain which goes under the fruit-wall, filling the air and the adjoining trees and shrubs with their numbers. The females were full of eggs. This late swarming is probably owing to the backward, wet season. The day following, not one flying ant was to be seen.

Horse-ants travel home to their nests laden with flies, which they have caught, and the aureliae of smaller ants, which they seize by violence. WHITE

In my *Naturalist's Calendar* for the year 1777, on September 6th, I find the following note to the article Flying Ants:

I saw a prodigious swarm of these ants flying about the top of some tall elm-trees (close by my house); some were continually dropping to the ground as if from the trees, and others rising up from the ground; many of them were joined together in copulation; and I imagine their life is but short, for as soon as produced from the egg by the heat of the sun, they propagate their species, and soon after perish. They were black, somewhat like the small black ant, and had four wings. I saw also, at another place, a large sort which were yellowish. On the eighth of September, 1785, I again observed the same circumstance of a vast number of these insects flying near the tops of the elms and dropping to the ground.

On the 2nd of March, 1777, I saw great numbers of ants come out of the ground. MARKWICK

Glow-worms

By observing two glow-worms which were brought into the field to the back in the garden, it appeared to us that these little creatures put out their lamps between eleven and twelve, and shine no more for the rest of the night.

Male glow-worms attracted by the light of the candles come into the parlour. WHITE

Earthworms

Earthworms make their casts most in mild weather about March and April; they do not lie torpid in winter, but come forth when there is no frost; they travel about in rainy nights, as appears from their sinuous tracks on the soft muddy soil, perhaps in search of food.

When earthworms lie out a-nights on the turf, though they extend their bodies a great way, they do not leave their holes, but keep the ends of their tails fixed therein, so that on the least alarm they can retire with precipitation under the earth. Whatever food falls within their reach when thus extended, they seem to be content with, such as blades of grass, straws, fallen leaves, the ends of which they often draw into their holes; even in copulation their hinder parts never quit their holes; so that no two, except they lie within reach of each other's bodies, can have any commerce of that kind; but as every individual is an hermaphrodite, there is no difficulty in meeting with a mate, as would be the case were they of different sexes. WHITE

Snails and Slugs

The shell-less snails called slugs are in motion all the winter in mild weather, and commit great depredations on garden plants, and much injure the green wheat, the loss of which is imputed to earthworms; while the shelled snail, the φερειοκος, does not come forth at all till about April 10th, and not only lays itself up pretty early in autumn, in places secure from frost, but also throws out round the mouth of its shell a thick operculum formed from its own saliva; so that it is perfectly secured, and corked up as it were, from all inclemencies. The cause why the slugs are able to endure the cold so much better than shell-snails is, that their bodies are covered with slime as whales are with blubber.

Snails copulate about Midsummer; and soon after deposit their eggs in the mould by running their heads and bodies under ground. Hence the way to be rid of them is to kill as many as possible before they begin to breed.

Larger grey, shell-less, cellar-snails lay themselves up about the same time with those that live abroad; hence it is plain that a defect of warmth is not the only cause that influences their retreat. WHITE

Snake's slough

. . . There the snake throws her enamell'd skin.
SHAKESPEARE'S *Midsummer Night's Dream*

About the middle of this month (September) we found in a field near a hedge the slough of a large snake, which seemed to have been newly cast. From circumstances it appeared as if turned wrong side outward, and as drawn off backward, like a stocking or woman's glove. Not only the whole skin, but scales from the very eyes, are peeled off, and appear in the head of the slough like a pair of spectacles. The reptile, at the time of changing his coat, had entangled himself intricately in the grass and weeds, so that the friction of the stalks and blades might promote this curious shifting of his exuviae.

. . . Lubrica serpens
Exuit in spinis vestem. LUCRETIUS

It would be a most entertaining sight could a person be an eyewitness to such a feat, and see the snake in the act of changing his garment. As the convexity of the scales of the eyes in the slough is now inward, that circumstance alone is a proof that the skin has been turned: not to mention that now the present inside is much darker than the outer. If you look through the scales of the snake's eyes from the concave side, viz. as the reptile used them, they lessen objects much. Thus it appears from what has been said, that snakes crawl out of the mouth of their own sloughs, and quit the tail part last, just as eels are skinned by a cook maid. While the scales of the eyes are growing loose, and a new skin is forming, the creature, in appearance, must be blind, and feel itself in an awkward, uneasy situation. WHITE

I have seen many sloughs or skins of snakes entire, after they have cast them off; and once in particular I remember to have found one of these sloughs so intricately interwoven amongst some brakes, that it was with difficulty removed without being broken: this undoubtedly was done by the creature to assist in getting rid of its incumbrance.

I have great reason to suppose that the eft or common lizard also casts its skin or slough, but not entire like the snake; for on the 30th of March, 1777, I saw one with something ragged hanging to it, which appeared to be part of its old skin. MARKWICK

Marelands

Observations on Vegetables

Trees, order of losing their leaves

One of the first trees that becomes naked is the walnut; the mulberry, the ash, especially if it bears many keys, and the horse-chestnut come next. All lopped trees, while their heads are young, carry their leaves a long while. Apple-trees and peaches remain green very late, often till the end of November: young beeches never cast their leaves till spring, till the new leaves sprout and push them off; in the autumn the beechen-leaves turn of a deep chestnut colour. Tall beeches cast their leaves about the end of October.

WHITE

Size and growth

Mr Marsham of Stratton, near Norwich, informs me by letter thus: 'I became a planter early; so that an oak which I planted in 1720 is become now, at one foot from the earth, twelve feet six inches in circumference, and at fourteen feet (the half of the timber length) is eight feet two inches. So if the bark was to be measured as timber, the tree gives 116½ feet, buyer's measure. Perhaps you never heard of a larger oak while the planter was living. I flatter myself that I increased the growth by washing the stem, and digging a circle as far as I supposed the roots to extend, and by spreading sawdust, &c., as related in the Phil. Trans. I wish I had begun with beeches (my favourite trees as well as yours), I might then have seen very large trees of my own raising. But I did not begin with beech till 1741, and then by seed; so that my largest is now at five feet from the ground, six feet three inches in girth, and with its head spreads a circle of twenty yards diameter. This tree was also dug round, washed, &c.' Stratton, 24th July, 1790.

The circumference of trees planted by myself at one foot from the ground (1790):

Oak	in 1730	4 feet 5 inches
Ash	in 1730	4 feet 6½ inches
Great fir	in 1751	5 feet
Greatest beech	in 1751	4 feet
Elm	in 1750	5 feet 3 inches
Lime	in 1756	5 feet 5 inches

The great oak in the Holt, which is deemed by Mr Marsham to be the biggest in this island, at seven feet from the ground, measures in circumference thirty-four feet. It has in old times lost several of its boughs, and is tending to decay. Mr Marsham computes, that at fourteen feet length this oak contains 1,000 feet of timber.

It has been the received opinion that trees grow in height only by their annual upper shoot. But my neighbour over the way, whose occupation confines him to one spot, assures me, that trees are expanded and raised in the lower parts also. The reason that he gives is this: the point of one of my firs began for the first time to peep over an opposite roof at the beginning of summer; but before the growing season was over, the whole shoot of the year, and three or four joints of the body beside, became visible to him as he sits on his form in his shop. According to this supposition, a tree may advance in height considerably, though the summer shoot should be destroyed every year. WHITE

Flowing of Sap

If the bough of a vine is cut late in the spring, just before the shoots push out, it will bleed considerably; but after the leaf is out, any part may be taken off without the least inconvenience. So oaks may be barked while the leaf is budding; but as soon as they are expanded, the bark will no longer part from the wood, because the sap that lubricates the bark and makes it part, is evaporated off through the leaves. WHITE

Renovation of Leaves

When oaks are quite stripped of their leaves by chaffers, they are clothed again soon after Midsummer with a beautiful foliage: but beeches, horse-chestnuts, and maples, once defaced by those insects, never recover their beauty again for the whole season. WHITE

Ash Trees

Many ash trees bear loads of keys every year, others never seem to bear any at all. The prolific ones are naked of leaves and unsightly; those that are sterile abound in foliage, and carry their verdure a long while, and are pleasing objects. WHITE

Beech

Beeches love to grow in crowded situations, and will insinuate themselves through the thickest covert, so as to surmount it all: they are therefore proper to mend thin places in tall hedges. WHITE

Sycamore

May 12. The sycamore or great maple is in bloom, and at this season makes a beautiful appearance, and affords much pabulum for bees, smelling strongly like honey. The foliage of this tree is very fine, and very ornamental to outlets. All the maples have saccharine juices. WHITE

Galls of Lombardy Poplar

The stalks and ribs of the leaves of the Lombardy poplar are embossed with large tumours of an oblong shape, which by incurious observers have been taken for the fruit of the tree. These galls are full of small insects, some of which are winged, and some not. The parent insect is of the genus of *cynips*. Some poplars in the garden are quite loaded with these excrescences. WHITE

Chestnut Timber

John Carpenter brings home some old chestnut trees which are very long; in several places the woodpeckers had begun to bore them. The timber and bark of these trees are so very like oak, as might easily deceive an indifferent observer, but the wood is very shaky, and towards the heart *cup-shaky* (that is to say, apt to separate in round pieces like cups), so that the inward parts are of no use. They are bought for the purpose of cooperage, but must make but ordinary barrels, buckets, &c. Chestnut sells for half the price of oak; but has sometimes been sent into the king's docks, and passed off instead of oak. WHITE

Lime Blossoms

Dr Chandler tells that in the south of France an infusion of the blossoms of the lime tree, *Tilia*, is in much esteem as a remedy for coughs, hoarsenesses, fevers, &c., and that at Nismes, he saw an avenue of limes that was quite ravaged and torn in pieces by people greedily gathering the bloom, which they dried and kept for these purposes.

Upon the strength of this information we made some tea of lime blossoms, and found it a very soft, well-flavoured, pleasant, saccharine julep, in taste much resembling the juice of liquorice. WHITE

Blackthorn

This tree usually blossoms while cold north-east winds blow; so that the harsh rugged weather obtaining at this season is called by the country people blackthorn winter. WHITE

Ivy Berries

Ivy berries form a noble and providential supply for birds in winter and spring; for the first severe frost freezes and spoils all the haws, sometimes by the middle of November; ivy berries do not seem to freeze. WHITE

Hops

The culture of Virgil's vines correspond very exactly with the modern management of hops. I might instance in the perpetual diggings and hoeings, in the tying to the stakes and poles, in pruning the superfluous shoots, &c., but lately I have observed a new circumstance, which was a neighbouring farmer's harrowing between the rows of hops with a small triangular harrow, drawn by one horse, and guided by two handles. This occurrence brought to my mind the following passage:

. . . ipsa
Flectere luctantes inter vineta juvencos.

VIRGIL's *Georgics*

Hops are dioecious plants; hence perhaps it might be proper, though not practised, to leave purposely some male plants in every garden, that their farina might impregnate the blossoms. The female plants without their male attendants are not in their natural state: hence we may suppose the frequent failure of crop so incident to hop-grounds; no other growth, cultivated by man, has such frequent and general failures as hops.

Two hop gardens much injured by a hailstorm, June 5, show now (September 2) a prodigious crop, and larger and fairer hops than any in the parish. The owners seem now to be convinced that the hail, by beating off the tops of the binds, has increased the side-shoots, and improved the crop. *Query.* Therefore should not the tops of hops be pinched off when the binds are very gross, and strong?

WHITE

Seed Lying Dormant

The naked part of the Hanger is now covered with thistles of various kinds. The seeds of these thistles may have lain probably under the thick shade of the beeches for many years, but could not vegetate till the sun and air were admitted. When old beech-trees are cleared away, the naked ground in a year or two becomes covered with strawberry plants, the seeds of which must have lain in the ground for an age at least. One of the slidders or trenches down the middle of the Hanger, close covered over with lofty beeches near a century old, is still called 'strawberry slidder', though no strawberries have grown there in the memory of man. That sort of fruit did once, no doubt, abound there, and will again when the obstruction is removed.

WHITE

Beans sown by Birds

Many horse-beans sprang up in my field-walks in the autumn, and are now grown to a considerable height. As the Ewel was in beans last summer, it is most likely that these seeds came from thence; but then the distance is too considerable for them to have been conveyed by mice. It is most probable therefore that they were brought by birds, and in particular by jays and pies, who seem to have hid them among the grass and moss, and then to have forgotten where they had stowed them. Some pease are growing also in the same situation, and probably under the same circumstances. WHITE

Cucumbers set by Bees

If bees, who are much the best setters of cucumbers, do not happen to take kindly to the frames, the best way is to tempt them by a little honey put on the male and female bloom. When they are once induced to haunt the frames, they set all the fruit, and will hover with impatience round the lights in a morning, till the glasses are opened. *Probatum est.* WHITE

Wheat

A notion has always obtained that in England hot summers are productive of fine crops of wheat; yet in the years 1780 and 1781, though the heat was intense, the wheat was much mildewed, and the crop light. Does not severe heat, while the straw is milky, occasion its juices to exude, which being extravasated, occasion spots, discolour the stems and blades, and injure the health of the plants? WHITE

Truffles

August. A truffle-hunter called on us, having in his pocket several large truffles found in this neighbourhood. He says these roots are not to be found in deep woods, but in narrow hedge-rows and the skirts of coppices. Some truffles, he informed us, lie two feet within the earth, and some quite on the surface; the latter, he added, have little or no smell, and are not so easily discovered by the dogs as those that lie deeper. Half-a-crown a pound was the price which he asked for this commodity. Truffles never abound in wet winters and springs. They are in season, in different situations, at least nine months in the year. WHITE

Tremella Nostoc.

Though the weather may have been ever so dry and burning, yet after two or three wet days this jelly-like substance abounds on the walks. WHITE

Fairy Rings

The cause, occasion, call it what you will, of fairy rings, subsists in the turf, and is conveyable with it: for the turf of my garden-walks, brought from the down above, abounds with those appearances, which vary their shape, and shift situation continually, discovering themselves now in circles, now in segments, and sometimes in irregular patches and spots. Wherever they obtain, puff-balls abound; the seeds of which were doubtless brought in the turf. WHITE

The Blacksmith's Shop

Meteorological Observations

Barometer

November 22, 1768. A remarkable fall of the barometer all over the kingdom. At Selborne we had no wind, and not much rain; only vast, swagging, rock-like clouds appeared at a distance. WHITE

Partial frost

The country people, who are abroad in winter mornings long before sunrise, talk much of hard frost in some spots, and none in others. The reason of these partial frosts is obvious, for there are at such times partial fogs about; where the fog obtains, little or no frost appears; but where the air is clear, there it freezes hard. So the frost takes place either on hill or in dale, wherever the air happens to be clearest and freest from vapour. WHITE

Thaw

Thaws are sometimes surprisingly quick, considering the small quantity of rain. Does not the warmth at such times come from below? The cold in still, severe seasons seems to come down from above; for the coming over of a cloud in severe nights raises the thermometer abroad at once full ten degrees. The first notices of thaws often seem to appear in vaults, cellars, &c.

If a frost happens, even when the ground is considerably dry, as soon as a thaw takes places, the paths and fields are all in a batter. Country people say that the frost draws moisture. But the true philosophy is, that the steam and vapours continually ascending from the earth, are bound in by the frost, and not suffered to escape till released by the thaw. No wonder then that the surface is all in a float; since the quantity of moisture by evaporation that arises daily from every acre of ground is astonishing. WHITE

Frozen Sleet

January 20. Mr H.'s man says that he caught this day in a lane near Hackwood park, many rooks, which, attempting to fly, fell from the trees with their wings frozen together by the sleet, that froze as it fell. There were, he affirms, many dozen so disabled. WHITE

Mist, called London Smoke

This is a blue mist which has somewhat the smell of coal smoke, and as it always comes to us with a north-east wind, is supposed to come from London. It has a strong smell, and is supposed to occasion blights. When such mists appear they are usually followed by dry weather. WHITE

Reflection on Fog

When people walk in a deep white fog by night with a lanthorn, if they will turn their backs to the light, they will see their shades impressed on the fog in rude gigantic proportions. This phenomenon seems not to have been attended to, but implies the great density of the meteor at that juncture. WHITE

Honey-dew

June 4, 1783. Fast honey-dews this week. The reason of these seem to be, that in hot days the effluvia of flowers are drawn up by a brisk evaporation, and then in the night fall down with the dews with which they are entangled.

This clammy substance is very grateful to bees, who gather it with great assiduity, but it is injurious to the trees on which it happens to fall, by stopping the pores of the leaves. The greatest quantity falls in still close weather; because winds disperse it, and copious dews dilute it, and prevent its ill effects. It falls mostly in hazy warm weather. WHITE

Morning Clouds

After a bright night and vast dew, the sky usually becomes cloudy by eleven or twelve o'clock in the forenoon, and clear again towards the decline of the day. The reason seems to be, that the dew, drawn up by evaporation, occasions the clouds; which, towards evening, being no longer rendered buoyant by the warmth of the sun, melt away, and fall down again in dews. If clouds are watched in a still warm evening, they will be seen to melt away and disappear. WHITE

Dripping Weather after Drought

No one that has not attended to such matters, and taken down remarks, can be aware how much ten days dripping weather will influence the growth of grass or corn after a severe dry season. This present summer, 1776, yielded a remarkable instance; for, till the 30th of May the fields were burnt up and naked, and the barley not half out of the ground; but now, June 10th, there is an agreeable prospect of plenty. WHITE

Aurora Borealis

November 1, 1787. The N. aurora made a particular appearance, forming itself into a broad, red, fiery belt, which extended from E. to W. across the welkin: but the moon rising at about ten o'clock in unclouded majesty in the E. put an end to this grand but awful meteorous phenomenon. WHITE

Black Spring, 1771

Dr Johnson says, that 'in 1771 the season was so severe in the island of Skye, that it is remembered by the name of the "black spring". The snow, which seldom lies at all, covered the ground for eight weeks, many cattle died, and those that survived were so emaciated that they did not require the male at the usual season.' The case was just the same with us here in the south; never were so many barren cows known as in the spring following that dreadful period. Whole dairies missed being in calf together.

At the end of March the face of the earth was naked to a surprising degree. Wheat hardly to be seen, and no signs of any grass; turnips all gone, and sheep in a starving way. All provisions rising in price. Farmers cannot sow for want of rain. WHITE

On the dark, still, dry, warm weather cccasionally happening in the winter months

Th' imprison'd winds slumber within their caves
Fast bound: the fickle vane, emblem of change,
Wavers no more, long settling to a point.
All nature nodding seems composed: thick stream
From land, from flood up-drawn, dimming the day
'Like a dark ceiling stand': slow thro' the air

Gossamer floats, or stretch'd from blade to blade
The wavy net-work whitens all the field.
  Push'd by the weightier atmosphere, up springs
The ponderous Mercury, from scale to scale
Mounting, amidst the Torricellian tube.[1]

  While high in air, and pois'd upon his wings
Unseen, the loft, enamour'd wood-lark runs
Thro' all his maze of melody; – the brake
  Loud with the black-bird's bolder note resounds.
Sooth'd by the genial warmth, the cawing rook
Anticipates the spring, selects her mate,
Haunts her tall nest-trees, and with sedulous care
Repairs her wicker eyrie, tempest torn.
  The ploughman inly smiles to see upturn
His mellow glebe, best pledge of future crop:
With glee the gardener eyes his smoking beds:
E'en pining sickness feels a short relief.

The happy schoolboy brings transported forth
His long forgotten scourge, and giddy gig:
O'er the white paths he whirls the rolling hoop,
Or triumphs in the dusty fields of taw.
Not so the museful sage: – abroad he walks
Contemplative, if haply he may find
What cause controls the tempest's rage, or whence
Amidst the savage season winter smiles.
For days, for weeks, prevails the placid calm.
At length some drops prelude a change: the sun
With ray refracted bursts the parting gloom;
When all the chequer'd sky is one bright glare.
Mutters the wind at eve: th' horizon round
With angry aspect scowls: down rush the showers,
And float the delug'd paths, and miry fields.

1 The Barometer.

Summary of the Weather

MEASURE OF RAIN IN INCHES

Year	Jan	Feb	Mar	April	May	June	July	Aug	Sept	Oct	Nov	Dec	Total
1782	4.64	1.98	6.54	4.57	6.34	1.75	7.09	8.28	3.72	1.93	2.51	0.91	50.26
1783	4.43	5.54	2.16	0.88	2.84	2.82	1.45	2.24	5.52	1.71	3.01	1.10	33.71
1784	3.18	0.77	3.82	3.92	1.52	3.65	2.40	3.88	2.51	0.39	4.70	3.06	33.80
1785	2.84	1.80	0.30	0.17	0.60	1.39	3.80	3.21	5.94	5.21	2.27	4.02	31.55
1786	6.91	1.42	1.62	1.81	2.40	1.20	1.99	4.34	4.79	5.04	4.38	—	—
1787	0.88	3.67	4.28	0.74	2.60	1.50	6.53	0.83	1.56	5.04	4.09	5.06	36.24
1788	1.60	3.37	1.31	0.61	0.76	1.27	3.58	3.22	5.71	0.0	0.86	0.23	22.50
1789	4.48	4.11	2.47	1.81	4.05	4.24	3.69	0.99	2.82	5.04	3.67	4.62	42.0
1790	1.99	0.49	0.45	3.64	4.38	0.13	3.24	2.30	0.66	2.10	6.95	5.94	32.27
1791	6.73	4.64	4.59	1.13	1.33	0.91	5.56	1.73	1.73	6.49	8.16	4.93	44.93
1792	6.07	1.68	6.70	4.08	3.00	2.78	5.16	4.25	5.53	5.55	1.65	2.11	48.56
1793	3.71	2.32	2.33	3.19	1.21								

1768 begins with a fortnight's frost and snow; rainy during February. Cold and wet spring; wet season from the beginning of June to the end of harvest. Latter end of September foggy without rain. All October and the first part of November rainy, and thence to the end of the year alternate rains and frosts.

1769. January and February, frosty and rainy, with gleams of fine weather in the intervals. To the middle of March, wind and rain. To the end of March, dry and windy. To the middle of April, stormy, with rain. To the end of June, fine weather, with rain. To the beginning of August, warm, dry weather. To the end of September, rainy with short intervals of fine weather. To the latter end of October, frosty mornings, with fine days. The next fortnight rainy; thence to the end of November dry and frosty. December, windy, with rain and intervals of frost, and the first fortnight very foggy.

1770. Frost for the first fortnight: during the 14th and 15th all the snow melted. To the end of February, mild hazy weather. The whole of March frosty, with bright weather. April, cloudy, with rain and snow. May began with summer showers, and ended with dark, cold rains. June, rainy, checquered with gleams of sunshine. The first fortnight in July, dark and sultry; the latter part of the month, heavy rain. August, September, and the first fortnight in October, in general, fine weather, though with frequent interruptions of rain; from the middle of October to the end of the year almost incessant rains.

1771. Severe frosts till the last week in January. To the first week in February, rain and snow: to the end of February, spring weather. To the end of the third week in April, frosty weather. To the end of the first fortnight in May, spring weather with copious showers. To the end of June, dry, warm weather. The first fortnight in July, warm, rainy weather. To the end of September, warm weather, but in general cloudy, with showers. October rainy. November frost, with intervals of fog and rain. December, in general, bright, mild weather, with hoar-frosts.

1772. To the end of the first week in February, frost and snow. To the end of the first fortnight in March, frost, sleet, rain, and snow. To the middle of April, cold rains. To the middle of May, dry weather, with cold piercing winds. To the end of the first week in June, cool showers. To the middle of August, hot, dry, summer weather. To the end of September, rain, with storms and thunder. To December 22, rain, with mild weather. December 23, the first ice. To the end of the month, cold, foggy weather.

1773. The first week in January, frost; thence to the end of the month, dark rainy weather. The first fortnight in February, hard frost. To the end of the first week in March, misty, showery weather. Bright spring days to the close of the month. Frequent showers to the latter end of April. To the end of June, warm showers with intervals of sunshine. To the end of August, dry weather with a few days of rain. To the end of the first fortnight in November, rainy. The next four weeks, frost: and thence to the end of the year, rainy.

1774. Frost and rain to the end of the first fortnight in March; thence to the end of the month, dry weather. To the 15th of April, showers; thence to the end of April, fine spring days. During May, showers and sunshine in about an equal proportion. Dark rainy weather to the end of the third week in July; thence to the 24th of August, sultry, with thunder and occasional showers. To the end of the third week in November, rain, with frequent intervals of sunny weather. To the end of December, dark, dripping fogs.

1775. To the end of the first fortnight in March, rain almost every day. To the first week in April, cold winds, with showers of rain and snow. To the end of June, warm, bright weather, with frequent showers. The first fortnight in July, almost incessant rains. To the 26th August, sultry weather with frequent showers. To the end of the third week in September, rain, with a few intervals of fine weather. To the end of the year, rain, with intervals of hoar-frost and sunshine.

1776. To January 24, dark frosty weather, with much snow. March 24, to the end of the month, foggy, with hoar-frost. To the 30th of May, dark, dry harsh weather, with cold winds. To the end of the first fortnight in July, warm, with much rain. To the end of the first week in August, hot and dry, with intervals of thunder showers. To the end of October, in general, fine seasonable weather, with a considerable proportion of rain. To the end of the year, dry, frosty weather, with some days of hard rain.

1777. To the 10th of January, hard frost. To the 20th of January, foggy, with frequent showers. To the 18th of February, hard dry frost with snow. To the end of May, heavy showers, with intervals of warm dry spring days. To the 8th July, dark with heavy rain. To the 18th July, dry, warm weather. To the end of July, very heavy rains. To the 12th October, remarkably fine warm weather. To the end of the year, grey mild weather with but little rain, and still less frost.

1778. To the 13th of January, frost, with a little snow: to the 24th January, rain: to the 30th, hard frost. To the 23rd February, dark, harsh foggy weather, with rain. To the end of the month, hard frost, with snow. To the end of the first fortnight in March, dark, harsh weather. From the first to the end of the first fortnight in April, spring weather. To the end of the month, snow and ice. To the 11th of June, cool with heavy showers. To the 19th July, hot, sultry, parching weather. To the end of the month, heavy showers. To the end of September, dry warm weather. To the end of the year wet, with considerable intervals of sunshine.

1779. Frost and showers to the end of January. To 21st April, warm dry weather. To 8th May, rainy. To 7th June, dry and warm. To the 6th July, hot weather, with frequent rain. To the 18th July, dry hot weather. To August 8, hot weather, with frequent rains. To the end of August, fine dry harvest weather. To the end of November, fine autumnal weather, with intervals of rain. To the end of the year, rain with frost and snow.

1780. To the end of January, frost. To the end of February, dark, harsh weather, with frequent intervals of frost. To the end of March, warm showery spring weather. To the end of April, dark, harsh weather, with rain and frost. To the end of the first fortnight in May, mild, with rain. To the end of August, rain and fair weather in pretty equal proportions. To the end of October, fine autumnal weather, with intervals of rain. To the 24th November, frost. To December 16, mild dry foggy weather. To the end of the year frost and snow.

1781. To January 25, frost and snow. To the end of February, harsh and windy, with rain and snow. To April 5, cold, drying winds. To the end of May, mild spring weather, with a few light showers. June began with heavy rain, but thence to the end of October, dry weather, with a few flying showers. To the end of the year, open weather with frequent rains.

1782. To February 4, open mild weather. To February 22, hard frost. To the end of March, cold blowing weather, with frost and snow and rain. To May 7, cold dark rains. To the end of May, mild, with incessant rains. To the end of June, warm and dry. To the end of August warm, with almost perpetual rains. The first fortnight in September mild and dry; thence to the end of the month, rain. To the end of October, mild with frequent showers. November began with hard frost, and continued throughout with alternate frost and thaw. The first part of December frosty: the latter part mild.

1783. To January 16, rainy with heavy winds. To the 24th, hard frost. To the end of the first fortnight in February, blowing, with much rain. To the end of February, stormy dripping weather. To the 9th of May, cold harsh winds (thick ice on 5th of May). To the end of August, hot weather, with frequent showers. To the 23rd September, mild, with heavy driving rains. To November 12, dry mild weather. To the 18th December, grey soft weather, with a few showers. To the end of the year hard frost.

1784. To February 19, hard frost, with two thaws; one the 14th January, the other 5th February. To February 28, mild wet fogs. To the 3rd March, frost with ice. To March 10, sleet and snow. To April 1, snow and hard frost. To April 27, mild weather, with much rain. To May 12, cold drying winds. To May 20, hot cloudless weather. To June 27, warm with frequent showers. To July 18, hot and dry. To the end of August, warm with heavy rains. To November 6, clear mild autumnal weather, except a few days of rain at the latter end of September. To the end of the year, fog, rain, and hard frost (on December 10, the thermometer 1 degree below 0).

1785. A thaw began on the 2nd January, and rainy weather with wind continued to January 28. To 15th March, very hard frost. To 21st March, mild, with sprinkling showers. To April 7, hard frost. To May 17, mild windy weather, without a drop of rain. To the end of May, cold, with a few showers. To June 9, mild weather, with frequent soft showers. To July 13, hot dry weather, with a few showery intervals. To July 22, heavy rain. To the end of September, warm with frequent showers. To the end of October frequent rain. To 18th of November, dry, mild weather. (Haymaking finished November 9, and the wheat harvest November 14.) To December 23, rain. To the end of the year, hard frost.

1786. To the 7th January, frost and snow. To January 13, mild with much rain. To 21st January, deep snow. To February 11, mild with frequent rains. To 21st February, dry, with high winds. To 10th March, hard frost. To 13th of April, wet, with intervals of frost. To the end of April, dry, mild weather. On the 1st and 2nd May, thick ice. To 10th May, heavy rain. To June 14, fine warm dry weather. From the 8th to the 11th July, heavy showers. To October 13, warm with frequent showers. To October 13, ice. To October 24, mild pleasant weather. To November 3, frost. To December 16, rain, with a few detached days of frost. To the end of the year, frost and snow.

1787. To January 24, dark, moist, mild weather. To January 28, frost and snow. To February 16, mild showery weather. To February 28, dry, cool weather. To March 10, stormy, with driving rain. To March 24, bright frosty weather. To the end of April, mild, with frequent rain. To May 22, fine bright weather. To the end of June, mostly warm, with frequent showers (on June 7, ice as thick as a crown piece). To the end of July, hot and sultry, with copious rain. To the end of September, hot dry weather, with occasional showers. To November 23, mild, with light frosts and rain. To the end of November, hard frost. To December 21, still and mild, with rain. To the end of the year, frost.

1788. To January 13, mild and wet. To January 18, frost. To the end of the month, dry, windy weather. To the end of February, frosty, with frequent showers. To March 14, hard frost. To the end of March, dark, harsh weather, with frequent showers. To April 4, windy, with showers. To the end of May, bright, dry, warm weather, with a few occasional showers. From June 28, to July 17, heavy rains. To August 12, hot dry weather. To the end of September, alternate showers and sunshine. To November 22, dry, cool weather. To the end of the year, hard frost.

1789. To January 13, hard frost. To the end of the month, mild, with showers. To the end of February, frequent rain, with snow-showers and heavy gales of wind. To 13th March, hard frost, with snow. To April 18, heavy rain, with frost and snow and sleet. To the end of April, dark, cold weather, with frequent rains. To June 9, warm spring weather, with brisk winds and frequent showers. From June 4 to the end of July, warm, with much rain. To August 29, hot, dry, sultry weather. To September 11, mild, with frequent showers. To the end of September, fine autumnal weather, with occasional showers. To November 17, heavy rain, with violent gales of wind. To December 18, mild, dry weather, with a few showers. To the end of the year, rain and wind.

1790. To January 16, mild, foggy weather, with occasional rains. To January 21, frost. To January 28, dark, with driving rains. To February 14, mild, dry weather. To February 22, hard frost. To April 5, bright cold weather, with a few showers. To April 15, dark and harsh, with a deep snow. To April 21, cold cloudy weather, with ice. To June 6, mild spring weather, with much rain. From July 3, to July 14, cool, with heavy rain. To the end of July, warm, dry weather. To August 6, cold, with wind and rain. To August 24, fine harvest weather. To September 5, strong gales, with driving showers. To November 26, mild autumnal weather, with frequent showers. To December 1, hard frost and snow. To the end of the year, rain and snow, and a few days of frost.

1791. To the end of January, mild, with heavy rains. To the end of February, windy, with much rain and snow. From March to the end of June, mostly dry, especially June. March and April rather cold and frosty. May and June, hot. July, rainy. Fine harvest weather, and pretty dry, to the end of September. Wet October, and cold towards the end. Very wet and stormy in November. Much frost in December.

1792. Some hard frost in January, but mostly wet and mild. February, some hard frost and a little snow. March, wet and cold. April, great storms on the 13th, then some very warm weather. May and June, cold and dry. July, wet and cool; indifferent harvest, rather late and wet. September, windy and wet. October, showery and mild. November, dry and fine. December, mild.

THE NATURALISTS' CALENDARS

At Oakhanger

A Comparative view of the Naturalists' Calendars as kept at Selborne in Hampshire by the late Revd Gilbert White, M.A. *and at Catsfield near Battle in Sussex by William Markwick, Esq.* F.L.S. *from the year 1768 to the year 1793*

The dates in the following calendars, when more than one, express the *earliest* and *latest* times in which the circumstances noted was observed.

Of the abbreviations used, fl. signifies flowering; lf. signifies leafing; and ap. the first appearance.

	WHITE	MARKWICK
Redbreast (*sylvia rubecula*) sings	Jan 1–12	Jan 3–31, and again Oct 6
Larks (*alauda arvensis*) congregate	Jan 1–18	Oct 16, Feb 9
Nuthatch (*sitta europaea*) heard	Jan 1–14	Mar 3, Apr 10
Winter aconite (*belleborus biemalis*) fl.	Jan 1, Feb 18	Feb 28, Apr 17
Shell-less snail or slug (*limax*) ap.	Jan 2	Jan 16, May 31
Grey wagtail (*motacilla boarula*) ap.	Jan 2–11	Jan 24, Mar 26
White wagtail (*motacilla alba*) ap.	Jan 2–11	Dec 12, Feb 23
Missel thrush (*turdus viscivorus*) sings	Jan 2–14	Feb 19, Apr 14
Bearsfoot (*belleborus foetidus*) fl.	Jan 2, Feb 14	Mar 1, May 5
Polyanthus (*primula polyantha*) fl.	Jan 2, Apr 12	Jan 1, Apr 9
Double daisy (*bellis perennis plena*) fl.	Jan 2, Feb 1	Mar 17, Apr 29
Mezereon (*daphne mezereum*) fl.	Jan 3, Feb 16	Jan 2, Apr 4
Pansy (*viola tricolor*) fl.	Jan 3	Jan 1, May 10
Red dead-nettle (*lamium purpureum*) fl.	Jan 3–21	Jan 1, Apr 5
Groundsel (*senecio vulgaris*) fl.	Jan 3–15	Jan 1, Apr 9
Hazel (*corylus avellana*) fl.	Jan 3, Feb 28	Jan 21, Mar 11

	WHITE	MARKWICK
Hepatica (*anemone hepatica*) fl.	Jan 4, Feb 18	Jan 17, Apr 9
Hedge-sparrow (*sylvia modularis*) sings	Jan 5–12	Jan 16, Mar 13
Common flies (*musca domestica*) seen in numbers	Jan 5, Feb 3	May 15
Greater titmouse (*parus major*) sings	Jan 6, Feb 6	Feb 17, Mar 17
Thrush (*turdus musicus*) sings	Jan 6–22	Jan 15, Apr 4
Insects swarm under sunny hedges	Jan 6	
Primrose (*primula vulgaris*) fl.	Jan 6, Apr 7	Jan 3, Mar 22
Bees (*apis mellifica*) ap.	Jan 6, Mar 19	Jan 31, Apr 11
Gnats play about	Jan 6, Feb 3	last seen Dec 30
Chaffinches, male and female (*fringilla coelebs*) seen in equal numbers)	Jan 6–11	Dec 2, Feb 3
Furze or gorse (*ulex europaeus*) fl.	Jan 8, Feb 1	Jan 1, Mar 27
Wall-flower (*cheiranthus cheiri*) *seu lfruticulosus* of Smith) fl.	Jan 8, Apr 1	Feb 21, May 9
Stock (*cheiranthus incanus*) fl.	Jan 8–12	Feb 1, June 3
Bunting (*Emberiza alba*) in great fl.ocks	Jan 9	
Linnets (*fringilla linota*) congregate	Jan 9	Jan 11
Lambs begin to fall	Jan 9–11	Jan 6, Feb 21
Rooks (*corvus frugilegus*) resort to their nest trees	Jan 10, Feb 11	Jan 23
Black hellebore (*helleborus niger*) fl.	Jan 10	Apr 27
Snowdrop (*galanthus nivalis*) fl.	Jan 10, Feb 5	Jan 18, Mar 1
White dead-nettle (*lamium album*) fl.	Jan 13	Mar 23, May 10
Trumpet honeysuckle, fl.	Jan 13	
Common creeping crowfoot (*ranunculus repens*) fl.	Jan 13	Apr 10, May 12
House-sparrow (*fringilla domestica*) chirps	Jan 14	Feb 17, May 9
Dandelion (*leontodon taraxacum*) fl.	Jan 16, Mar 11	Feb 1, Apr 17
Bat (*vespertilio*) ap.	Jan 16, Mar 24	Feb 6, June 1
Spiders shoot their webs	Jan 16	last seen Nov20
Butterfly ap.	Jan 16	Feb 21, May 8 last seen Dec 22
Brambling (*fringilla montifringilla*) ap.	Jan 16	Jan 10 –31
Blackbird (*turdus merula*) whistles	Jan 17	Feb 15, May 13
Wren (*sylvia troglodytes*) sings	Jan 17	Feb 7, June 12
Earthworms lie out	Jan 18, Feb 8	
Crocus (*crocus vernus*) fl.	Jan 13, Mar 18	Jan 20, Mar 19

	WHITE	MARKWICK
Skylark (*alauda arvensis*) sings	Jan 21	Jan 12, Feb 27 sings till Nov 13
Ivy casts its leaves	Jan 22	
Helleborus hiemalis fl.	Jan 22–24	Feb 28, Apr 17
Common dor or clock (*scarabaeus stercorarius*)	Jan 23	Feb 12, Apr 12 last seen Nov 24
Peziza acetabulum ap.	Jan 23	
Helleborus virid fl..	Jan 23, Mar 5	
Hazel (*corylus avellana*) fl..	Jan 23, Feb 1	Jan 27, Mar 11
Woodlark (*alauda arborea*) sings	Jan 24, Feb 21	Jan 28, June 5
Chaffinch (*fringilla coelebs*) sings	Jan 24, Feb 15	Jan 21, Feb 26
Jackdaws begin to come to churches	Jan 25, Mar 4	last seen Sept 8
Yellow wagtail (*motacilla flava*) ap.	Jan 25, Apr 14	Apr 13, July 3
Honeysuckle (*lonicera periclymenum*) lf.	Jan 25	Jan 1, Apr 9
Field or procumbent speedwell (*veronica agrestis*) fl.	Jan27, Mar 15	Feb 12, Mar 29
Nettle butterfly (*papilio urticae*) ap.	Jan 27, Apr 2	Mar 5, Apr 24 last seen June 6
White wagtail (*motacilla alba*) chirps	Jan 28	Mar 16
Shell-snail (*helix nemoralis*) ap.	Jan 28, Feb 24	Apr 2, June 11
Earthworms engender	Jan 30	
Barren strawberry (*fragaria sterilis*) fl.	Feb 1, Mar 26	Jan 13, Mar 26
Blue titmouse (*parus caeruleus*) chirps	Feb 1	Apr 27
Brown wood-owls hoot	Feb 2	
Hen (*phasianus gallus*) sits	Feb 3	Mar 8, hatches
Marsh titmouse begins his two harsh sharp notes	Feb 3	
Gossamer floats	Feb 4, Apr 1	
Musca tenax ap.	Feb 4, Apr 8	
Laurustine (*viburnum tinus*) fl..	Feb 5	Jan 1, Apr 5
Butcher's broom (*ruscus aculeatus*) fl.	Feb 5	Jan 1, May 10
Fox (*canis vulpes*) smells rank	Feb 7	May 19 young brought forth
Turkey-cocks strut and gobble	Feb 10	
Yellow-hammer (*emberiza citrinella*) sings	Feb 12	Feb 18, Apr 28
Brimstone butterfly (*papilio rhamni*) ap.	Feb 13, Apr 2	Feb 13, Mar 8 last seen Dec 24
Green-woodpecker (*picus viridis*) makes a loud cry	Feb 13, Mar 23	Jan 1, Apr 17

	WHITE	MARKWICK
Raven (*corvus corax*) builds	Feb 14–17	Apr 1, has young ones June 1
Yew-tree (*taxus baccata*) fl.	Feb 14, Mar 27	Feb 2, Apr 11
Coltsfoot (*tussilago farfara*) fl.	Feb 15, Mar 23	Feb 18, Apr 13
Rooks (*corvus frugilegus*) build	Feb 16, Mar 6	Feb 28, Mar 5
Partridges (*perdix cinerea*) pair	Feb 17	Feb 16, Mar 20
Peas (*pisum sativum*) sown	Feb 17, Mar 8	Feb 8, Mar 31
House-pigeon (*columba domestica*) has young ones	Feb 18	Feb 8
Field-crickets open their holes	Feb 20, Mar 30	
Common flea (*pulex irritans*) ap.	Feb 21–26	
Pilewort (*ficaria verna*) fl.	Feb 21, Apr 13	Jan 25, Mar 26
Goldfinch (*fringilla carduelis*) sings	Feb 21, Apr 5	Feb 28, May 5
Viper (*coluber berus*) ap.	Feb 22, Mar 26	Feb 23, May 6 last seen Oct 28
Wood-louse (*oniscus asellus*) ap.	Feb 23, Apr 1	Apr 27, June 17
Missel thrushes pair	Feb 24	
Daffodil (*narcissus pseudonarcissus*) fl.	Feb 24, Apr 7	Feb 26, Apr 18
Willow (*salix alba*) fl.	Feb 24, Apr 2	Feb 27, Apr 11
Frogs (*rana temporaria*) croak	Feb 25	Mar 9, Apr 20
Sweet violet (*viola odorata*) fl.	Feb 26, Mar 31	Feb 7, Apr 5
Phalaena tinea vestianella ap.	Feb 26	
Stone-curlew (*otis oedicnemus*) clamours	Feb 27, Apr 24	June 17
Filbert (*corylus sativus*) fl.	Feb 27	Jan 25, Mar 26
Ring-dove cooes	Feb 27, Apr 5	Mar 2, Aug 10
Apricot-tree (*prunus armeniaca*) fl.	Feb	Feb 28, Apr 5
Toad (*rana bufo*) ap.	Feb 28, Mar 24	Mar 15, July 1
Frogs (*rana temporaria*) spawn	Feb 28, Mar 22	Feb 9, Apr 10, tadpoles Mar 19
Ivy-leaved speedwell (*veronica hederifolia*) fl.	Mar 1, Apr 2	Feb 16, Apr 10
Peach (*amygdalus persica*) fl.	Mar 2, Apr 17	Mar 4, Apr 29
Frog (*rana temporaria*) ap.	Mar 2, Apr 6	Mar 9
Shepherd's purse (*thlaspi bursapastoris*) fl.	Mar 3	Jan 2, Apr 16
Pheasant (*phasianus colchicus*) crows	Mar 3–29	Mar 1, May 22
Land-tortoise comes forth	Mar 4, May 8	
Lungwort (*pulmonaria officinalis*) fl.	Mar 4, Apr 16	Mar 2, May 19
Podura fimetaria ap.	Mar 4	
Aranea scenica saliens ap.	Mar 4	

	WHITE	MARKWICK
Scolopendra forficata ap.	Mar 5–16	
Wryneck (*jynx torquilla*) ap.	Mar 5, Apr 25	Mar 26, Apr–23, last seen Sept 14
Goose (*anas anser*) sits on its eggs	Mar 5	Mar 21
Duck (*anas boschas*) lays	Mar 5	Mar 28
Dog's violet (*viola canina*) fl.	Mar 6, Apr 18	Feb 28, Apr 22
Peacock butterfly (*papilio io*) ap.	Mar 6	Feb 13, Apr 20 last seen Dec 25
Trouts begin to rise	Mar 7–14	
Field beans (*vicia faba*) planted	Mar 8	Apr 26, emerge
Blood-worms appear in the water	Mar 8	
Crow (*corvus corone*) builds	Mar 10	July 1 has young ones
Oats (*avena sativa*) sown	Mar 10–18	Mar 16, Apr 13
Golden-crowned wren (*sylvia regulus*) sings	Mar 12, Apr 30	Apr 15, May 22, seen Dec 23, Jan 26
Asp (*populus tremula*) fl.	Mar 12	Feb 26, Mar 28
Common elder (*sambucus nigra*) lf.	Mar 13–20	Jan 24, Apr 22
Laurel (*prunus laurocerasus*) fl.	Mar 15, May 21	Apr 2, May 27
Chrysomela Gotting ap.	Mar 15	
Black ants (*formica nigra*) ap.	Mar 15, Apr 22	Mar 2, May 18
Ephemerae bisetae ap.	Mar 16	
Gooseberry (*ribes grossularia*) lf.	Mar 17, Apr 11	Feb 26, Apr 9
Common stitchwort (*stellaria holostea*)	Mar 17, May 19	Mar 8, May 7
Wood anemone (*anemone nemerosa*) fl.	Mar 17, Apr 22	Feb 27, Apr 10
Blackbird (*turdus merula*) lays	Mar 17	Apr 14 young ones May 19
Raven (*corvus corax*) sits	Mar 17	Apr 1, builds
Wheatear (*sylvia oenanthe*) ap.	Mar 18–30	Mar 13, May 23, last seen Oct 26
Musk-wood crowfoot (*adoxa moschatellina*) fl.	Mar 18, Apr 13	Feb 23, Apr 28
Willow-wren (*sylvia trochilus*) ap.	Mar 19, Apr 13	Mar 30, May 16 sits May 27 last seen Oct 23
Fumaria bulboso fl.	Mar 19	
Elm (*ulmus campestris*) fl.	Mar 19, Apr 4	Feb 17, Apr 25
Turkey (*meleagris gallopavo*) lays	Mar 19, Apr 7	Mar 18–25 sits Apr 4 young ones Apr 30

	WHITE	MARKWICK
House pigeons (*columba domestica*) sit	Mar 20	Mar 20 young hatched
Marsh marigold (*caltha palustris*) fl.	Mar 20, Apr 14	Mar 22, May 8
Buzz-fly (*bombylius medius*) ap.	Mar 21, Apr 28	Mar 15, Apr 30
Sand-martin (*hirundo riparia*) ap.	Mar 21, Apr 12	Apr 8, May 16 last seen Sept 8
Snake (*coluber natrix*) ap.	Mar 22–30	Mar 3, Apr 29 last seen Oct 2
Horse ant (*formica herculeana*) ap.	Mar 22, Apr 18	Feb 4, Mar 26 last seen Nov 12
Greenfinch (*loxia chloris*) sings	Mar 22, Apr 22	Mar 6, Apr 26
Ivy (*hedera helix*) berries ripe	Mar 23, Apr 14	Feb 16, May 19
Periwinkle (*vinca minor*) fl.	Mar 25	Feb 6, May 7
Spurge laurel (*daphne laureola*) fl.	Mar 25, Apr 1	Apr 12–22
Swallow (*hirundo rustica*) ap.	Mar 26, Apr 20	Apr 7–27 last seen Nov 16
Black-cap (*sylvia atricapilla*) heard	Mar 26, May 4	Apr 14, May 18, seen Apr 14, May 20, last seen Sept 19
Young ducks hatched	Mar 27	Apr 6, May 16
Golden saxifrage (*chrysosplenium oppositifolium*) fl.	Mar 27, Apr 9	Feb 7, Mar 27
Martin (*hirundo urbica*) ap.	Mar 28, May 1	Apr 14, May 8 last seen Dec 8
Double hyacinth (*hyacinthus orientalis*) fl.	Mar 29, Apr 22	Mar 13, Apr 24
Young geese (*anas anser*)	Mar 29	Mar 29, Apr 19
Wood sorrel (*oxalis acetosella*) fl.	Mar 30, Apr 22	Feb 26, Apr 26
Ring-ousel (*turdus torquatus*) seen	Mar 30, Apr 17	Oct 11
Barley (*hordeum sativum*) sown	Mar 31, Apr 30	Apr 12, May 20
Nightingale (*sylvia luscinia*) sings	Apr 1, May 1	Apr 5, July 4 last seen Aug29
Ash (*fraxinus excelsior*) fl.	Apr 1, May 4	Mar 16, May 8
Spiders' webs on the surface of the ground	Apr 1	
Chequered daffodil (*fritillaria meleagris*) fl.	Apr 2–24	Apr 15, May 1
Julus terrestris ap.	Apr 2	
Cowslip (*primula veris*) fl.	Apr 3–24	Mar 3, May 17
Ground-ivy (*glecoma hederacea*) fl.	Apr 3–15	Mar 2, Apr 16
Snipe pipes	Apr 3	

	WHITE	MARKWICK
Box-tree (*buxus sempervirens*) fl.	Apr 3	Mar 27, May 8
Elm (*ulmus campestris*) lf.	Apr 3	Apr 2, May 19
Gooseberry (*ribes grossularia*) fl.	Apr 3–14	Mar 21, May 1
Currant (*ribes hortense*) fl.	Apr 3–5	Mar 24, Apr 28
Pear-tree (*pyrus communis*) fl.	Apr 3, May 29	Mar 30, Apr 30
Newt or eft (*Lacerta vulgaris*)	Apr 4	Feb 17, Apr 15 last seen Oct 9
Dogs' mercury (*mercurialis perennis*) fl.	Apr 5–19	Jan 20, Apr 16
Wych elm (*ulmus glabra seu montana* of Smith) fl.	Apr 5	Apr 19, May 10
Ladysmock (*cardamine pratensis*) fl.	Apr 6–20	Feb 21, Apr 26
Cuckoo (*cuculus canorus*) heard	Apr 7–26	Apr 15, May 3 last heard June 28
Blackthorn (*prunus spinosa*) fl.	Apr May 10	Mar 16, May 8
Death-watch (*termes pulsatorius*) beats	Apr 7	Mar, 28, May 28
Gudgeon spawns	Apr 7	
Red-start (*sylvia phaenicurus*) ap.	Apr 8–28	Apr 5, sings Apr 15, last seen Sept 30
Crown imperial (*fritillaria imperialis*) fl.	Apr 8–24	Apr 1, May 13
Tit-lark (*alauda pratensis*) sings	Apr 9–19	Apr 14–29 sits June 16–17
Beech (*fagus sylvatica*) lf.	Apr 10, May 8	Apr 24, May 25
Shell-snail (*helix memoralis*) comes out in troops	Apr 11, May 9	May 17, June 11 ap.
Middle yellow wren, ap.	Apr 11	
Swift (*hirundo apus*) ap.	Apr 13, May 7	Apr 28, May 19
Stinging-fly (*conops calcitrans*) ap.	Apr 14, May 17	
Whitlow grass (*draba verna*) fl.	Apr 14	Jan 15, Mar 24
Larch-tree (*pinus-larix rubra*) lf.	Apr 14	Apr 1, May 9
Whitethroat (*sylvia cinerea*) ap.	Apr 14, May 14	Apr 14, May 5 sings May 3–10 last seen Sept 23
Red ant (*formica rubra*) ap.	Apr 14	Apr 9, June 26
Mole cricket (*gryllus gryllotalpa*) churs.	Apr 14	
Second willow or laughing wren, ap.	Apr 14–19–23	
Red rattle (*pedicularis sylvatica*) fl.	Apr 15–19	Apr 10, June 4
Common flesh-fly (*musca carnaria*) ap.	Apr 15	
Lady-cow (*coccinella bipunctata*) ap.	Apr 16	
Grasshopper lark (*alauda locustae voce*) ap.	Apr 16–30	

	WHITE	MARKWICK
Willow-wren, its shivering note heard	Apr 17, May 7	Apr 28, May 14
Middle willow-wren (*regulus non cristatus-medius*) ap.	Apr 17, May 2	
Wild cherry (*prunus cerasus*) fl.	Apr 18, May 12	Mar 30, May 10
Garden cherry (*prunus cerasus*) fl.	Apr 18, May 11	Mar 25, May 6
Plum (*prunus domestica*) fl.	Apr 18, May 5	Mar 24, May 6
Harebell (*hyacinthus non scriptus seu scilla nutans* of Smith) fl.	Apr 19–25	Mar 27, May 8
Turtle (*columba turta*) cooes	Apr 20–27	May 14, Aug 10
Hawthorn (*crataegus seu mespilus oxycantha* of Smith) fl.	Apr 20, June 11	Apr 19, May 26
Male fool's orchis (*orchis mascula*) fl.	Apr 21	Mar 29, May 13
Blue flesh-fly (*musca vomitoria*) ap.	Apr 21, May 23	
Black snail or slug (*limax ater*) abound	Apr 22	Feb 1, Oct 24, ap.
Apple-tree (*pyrus malus sativus*) fl.	Apr 22, May 25	Apr 11, May 26
Large bat, ap.	Apr 22, June 11	
Strawberry, wild wood (*fragaria vesca sylv.*) fl.	Apr 23–29	Apr 8–9
Sauce alone (*erysimum alliaria*) fl.	Apr 23	Mar 31, May 8
Wild or bird cherry (*prunus avium*) fl.	Apr 24	Mar 30, May 10
Apis hypnorum, ap.	Apr 24	
Musca meridiana, ap.	Apr 24, May 28	
Wolf-fly (*asilus*) ap.	Apr 25	
Cabbage-butterfly (*papilio brassicae*) ap.	Apr 28, May 20	Apr 29, June 10
Dragonfly (*libellula*) ap.	Apr 30, May 21	Apr 18, May 13 last seen Nov 10
Sycamore (*acer pseudoplatanus*) fl.	Apr 30, June 6	Apr 20, June 4
Bombylus minor, ap.	May 1	
Glow-worm (*lampyris noticula*) shines	May 1, June 11	June 19, Sept 28
Fern-owl or goatsucker (*caprimulgus europaeus*) ap.	May 1-26	May 16, Sept 14
Common bugle (*ajuga reptans*) fl.	May 1	Mar 27, May 10
Field-crickets (*gryllus compestris*) crink	May 2–24	
Chaffer or May-bug (*scarabaeus melolontha*) ap.	May 2–26	May 2, July 7
Honeysuckle (*lonicera periclymenum*) fl.	May 3–30	Apr 24, June 21
Toothwort (*lathraea squamaria*) fl.	May 4–12	
Shell-snails copulate	May 4, June 17	
Sedge warbler (*sylvia salicaria*) sings	May 4	June 2–30

	WHITE	MARKWICK
Mealy tree (*viburnum lantana*) fl.	May 5–17	Apr 25, May 2
Fly-catcher (*stoparolas muscicapa grisola*) ap.	May 10–30	Apr 29, May 21
Apis longicornis, ap.	May 10, June 9	
Sedge warbler (*sylvia salicaria*) ap.	May 11–13	Aug 2
Oak (*quercus robur*) fl.	May 13–15	Apr 29, June 4
Admiral-butterfly (*papilio atalanta*) ap.	May 13	
Orange-tip (*papilio cardamines*) ap.	May 14	Mar 30, May 19
Beech (*fagus sylvatica*) fl.	May 15–26	Apr 23, May 28
Common maple (*acer campestre*) fl.	May 16	Apr 24, May 27
Barberry tree (*berberis vulgaris*) fl.	May 17–26	Apr 28, June 4
Wood argus-butterfly (*papilio aegeria*) ap.	May 17	
Orange lily (*lilium bulbiferum*) fl.	May 18, June 11	June 14, July 22
Burnet-moth (*sphinx filipendulae*) ap.	May 18, June 13	May 24, June 26
Walnut (*juglans regia*) lf.	May 18	Apr 10 June 1
Laburnum (*cytisus laburnum*) fl.	May 18, June 5	May 1, June 23
Forest-fly (*hippobosca equina*) ap.	May 18, June 9	
Saintfoin (*hedysarum onobrychis*) fl.	May 19, June 8	May 21, July 28
Peony (*paeonia officinalis*) fl.	May 20, June 15	Apr 18, May 26
Horse chestnut (*aesculus hippocastanum*) fl.	May 21, June 9	Apr 19, June 7
Lilac (*syringa vulgaris*) fl.	May 21	Apr 15, May 30
Columbine (*aquilegia vulgaris*) fl.	May 21-27	May 6, June 13
Medlar (*mespilus germanica*) fl.	May 21, June 20	Apr 8, June 19
Tormentil (*tormentilla erecta seu officinalis* of Smith) fl.	May 21	Apr 17, June 11
Lily of the valley (*convallaria majalis*) fl.	May 22	Apr 27, June 13
Bees (*apis mellifica*) swarm	May 22, July 22	May 12, June 23
Woodroof (*asperula odorata*) fl.	May 22–25	Apr 14, June 4
Wasp, female (*vespa vulgaris*) ap.	May 23	Apr 2, June 4 last seen Nov 2
Mountain ash (*sorbus seu pyrus aucuparia* of Smith) fl.	May 23, June 8	Apr 20, June 8
Birds'-nest orchis (*ophrys nidus avis*) fl.	May 24, June 11	May 18, June 12
White-beam tree (*crataegus seu pyrus aria* of Smith) lf.	May 24, June 4	May 3
Milkwort (*polygala vulgaris*) fl.	May 24, June 7	Apr 13, June 2
Dwarf cistus (*cistus helianthemum*) fl.	May 25	May 4, Aug 8
Gelder rose (*viburnum opulus*) fl.	May 26	May 10, June 8
Common elder (*sambucus nigra*) fl.	May 26, June 25	May 6, June 17

	WHITE	MARKWICK
Cantharis noctiluca, ap.	May 26	
Apis longicornis bores holes in walls	May 27, June 9	
Mulberry tree (*morus nigra*) lf.	May 27, June 13	May 20, June 11
Wild-service tree (*crataegus seu pyrus torminalis* of Smith) fl.	May 27	May 13, June 19
Sanicle (*sanicula europaea*) fl.	May 27, June 13	Apr 23, June 4
Avens (*geum urbanum*) fl.	May 28	May 9, June 11
Female fool's orchis (*orchis morio*) fl.	May 28	Apr 17, May 20
Ragged Robin (*lychnis flos cuculi*) fl.	May 29, June 1	May 12, June 8
Burnet (*poterium sanguisorba*) fl.	May 29	Apr 30, Aug 7
Foxglove (*digitalis purpurea*) fl.	May 30, June 22	May 23, June 15
Corn-flag (*gladiolus communis*) fl.	May 30, June 20	June 9, July 8
Serapias longifol, fl.	May 30, June 13	
Raspberry (*rubus idaeus*) fl.	May 30, June 31	May 10, June 16
Herb Robert (*geranium Robertianum*) fl.	May 30	Mar 7, May 16
Figwort (*scrophularia nodosa*) fl.	May 31	May 12, June 20
Gromwell (*lithospermum officinale*) fl.	May 31	May 10–24
Wood spurge (*euphorbia amygdaloides*) fl.	June 1	Mar 23, May 13
Ramsons (*allium ursinum*) fl.	June 1	Apr 21, June 4
Mouse-ear scorpion grass (*myosotis scorpiodes*) fl.	June 1	Apr 11, June 1
Grasshopper (*gryllus grossus*) ap.	June 1-14	Mar 25, July 6 last seen Nov 3
Rose (*rosa hortensis*) fl.	June 1–21	June 7, July 1
Mouse-ear hawkweed (*hieracium pilosella*) fl.	June 1, July 16	Apr 19, June 12
Buckbean (*menyanthes trifoliata*) fl.	June 1	Apr 20, June 8
Rose-chaffer (*scarabaeus auratus*) ap.	June 2–8	Apr 18, Aug 4
Sheep (*ovis aries*) shorn	June 2–23	May 23, June 17
Water-flag (*iris pseudo-acorus*) fl.	June 2	May 8, June 9
Cultivated rye (*secale cereale*) fl.	June 2	May 27
Hounds' tongue(*cynoglossum officinale*) fl.	June 2	May 11, June 7
Helleborine (*serapias latifolia*) fl.	June 2, Aug 6	July 22, Sept 6
Green-gold fly (*musca caesar*) ap.	June 2	
Argus butterfly (*papilio moera*) ap.	June 2	
Spearwort (*ranunculus flammula*) fl.	June 3	Apr 25, June 13
Birdsfoot trefoil (*lotus corniculatus*) fl.	June 3	Apr 10, June 3
Fraxinella or white dittany (*dictamnus albus*) fl.	June 3–11	June 9, July 24

	WHITE	MARKWICK
Phryganea nigra, ap.	June 3	
Angler's May-fly (*ephemera vulg*) ap.	June 3–14	
Lady's finger (*anthyllis vulneraria*) fl.	June 4	June 1, Aug 16
Bee-orchis (*ophrys apifera*) fl.	June 4, July 4	
Pink (*dianthus deltoides*) fl.	June 5–19	May 26, July 6
Mock orange (*philadelphus coronarius*) fl.	June 5	May 16, June 23
Libellula virgo, ap.	June 5–20	
Vine (*vitus vinifera*) fl.	June 7, July 30	June 18, July 29
Portugal laurel (*prunus lusitanicus*) fl.	June 8, July 1	June 3, July 16
Purple-spotted martagon (*lillum martagon*) fl.	June 8–25	June 18, July 19
Meadow cranes-bill (*geranium pratense*) fl.	June 8, Aug 1	
Black bryony (*tamus communis*) fl.	June 8	May 15, June 21
Field pea (*pisum sativum arvense*) fl.	June 9	May 15, June 21
Bladder campion (*cucubalus behen seu silene inflata* of Smith) fl.	June 9	May 4, July 13
Bryony (*brionia alba*) fl.	June 9	May 13, Aug 17
Hedge-nettle (*stachys sylvatica*) fl.	June 10	May 28, June 24
Bittersweet (*solanum dulcamara*) fl.	June 11	May 15, June 20
Walnut (*juglans regia*) fl.	June 12	Apr 18, June 1
Phallus impudicus, ap.	June 12, July 23	
Rosebay willow-herb (*epilobium angustifolium*) fl.	June 12	June 4, July 28
Wheat (*triticum hybernum*) fl.	June 13, July 22	June 4–30
Comfrey (*symphytum officinale*) fl.	June 13	May 4, June 23
Yellow pimpernel (*lysimachia nemorum*) fl.	June 13–30	Apr 10, June 12
Tremella nostoc, ap.	June 15, Aug 24	
Buckthorn (*rhamnus catharticus*) lf.	June 16	May 25
Cuckow-spit insect (*cicadia spumaria*) ap.	June 16	June 2–21
Dog-rose (*rosa canina*) fl.	June 17, 18	May 24, June 21
Puff-ball (*lycoperdon bovista*) ap.	June 17, Sept 3	May 6, Aug 19
Mullein (*verbascum thapsus*) fl.	June 18	June 10, July 22
Viper's bugloss (*echium anglicum seu vulgare* of Smith) fl.	June 19	May 27, July 3
Meadow hay cut	June 19, July 20	June 13, July 7
Stag-beetle (*lucanus cervus*) ap.	June 19	June 14–21
Borage (*borago officinalis*) fl.	June 20	Apr 22, July 26
Spindle-tree (*euonymus europaeus*) fl.	June 20	May 11, June 25

	WHITE	MARKWICK
Musk thistle (*carduus nutans*) fl.	June 20, July 4	June 4, July 25
Dogwood (*cornus sanguinea*) fl.	June 21	May 28, June 27
Field scabious (*scabiosa arvensis*) fl.	June 21	June 16, Aug 14
Marsh thistle (*carduus palustris*) fl.	June 21–27	May 15, June 19
Dropwort (*spiraea filipendula*) fl.	June 22, July 9	May 8, Sept 3
Great wild valerian (*valeriana officinalis*) fl.	June 22, July 7	May 22, July 21
Quail (*perdix coturnix*) calls	June 22, July 4	July 23 seen Sept 1–18
Mountain willow-herb (*epilobium montanum*) fl.	June 22	June 5–21
Thistle upon thistle (*carduus crispus*) fl.	June 23–29	May 22, July 22
Cow-parsnip (*heracleum sphondylium*)fl	June 23	May 27, July 12
Earth-nut (*bunium bulbocastanum seu flexuosum* of Smith) fl.	June 23	May 4–31
Young frogs migrate	June 23, Aug 2	
Oestrus curvicauda, ap.	June 24	
Vervain (*verbena officinalis*) fl.	June 24	June 10, July 17
Corn poppy (*papaver rhoeas*) fl.	June 24	Apr 30, July 15
Self-heal (*prunella vulgaris*) fl.	June 24	June 7–23
Agrimony (*agrimonia eupatoria*) fl.	June 24–29	June 7, July 9
Great horse-fly (*tabanus bovinus*) ap.	June 24, Aug 2	
Greater knapweed (*centaurea scabiosa*) fl.	June 25	June 7, Aug 14
Mushroom (*agaricus campestris*) ap.	June 26, Aug 30	Apr 16, Aug 16
Common mallow (*malva sylvestris*) fl.	June 26	May 27, July 13
Dwarf mallow (*malva rotundifolia*) fl.	June 26	May 12, July 20
St John's wort (*hypericum perforatum*) fl.	June 26	June 15, July 12
Broom-rape (*orobanche major*) fl.	June 27, July 4	May 9, July 25
Henbane (*hyoscyamus niger*) fl.	June 27	May 13, June 19
Goat's-beard (*tragopogon pratense*) fl.	June 27	June 5–14
Deadly nightshade (*atropa belladonna*) fl.	June 27	May 22, Aug 14
Truffles begin to be found	June 28, July 29	
Young partridges fly	June 28, July 31	July 8–28
Lime-tree (*tilia europaea*) fl.	June 28, July 31	June 12, July 30
Spearthistle (*carduus lanceolatus*) fl.	June 28, July 12	June 27, July 18
Meadow-sweet (*spiraea ulmaria*) fl.	June 28	June 16, July 24
Greenweed (*genista tinctoria*) fl.	June 28	June 4, July 24
Wild thyme (*thymus serpyllum*) fl.	June 28	June 6, July 19
Stachys germanic, fl.	June 29, July 20	

	WHITE	MARKWICK
Day-lily (*hemerocallis flava*) fl.	June 29, July 4	May 29, June 9
Jasmine (*jasminum officinale*) fl.	June 29, July 30	June 27, July 21
Holly-oak (*alcea rosea*) fl.	June 29, Aug 4	July 4, Sept 7
Monotropa hypopithys, fl.	June 29, July 23	
Ladies' bedstraw (*galium verum*) fl.	June 29	June 22, Aug 3
Galium palustre, fl.	June 29	
Nipplewort (*lapsana communis*) fl.	June 29	May 30, July 24
Welted thistle (*carduus acanthoides*) fl.	June 29	
Sneezewort (*achillea ptarmica*) fl.	June 30	June 22, Aug 3
Musk mallow (*malva moschata*) fl.	June 30	June 9, July 14
Pimpernel (*analgallis arvensis*) fl.	June 30	May 4, June 22
Hoary-beetle (*scarabaeus solstit*) ap.	June 30, July 17	
Corn saw-wort (*serratula arvensis seu carduus arvensis* of Smith) fl.	July 1	June 15, July 15
Pheasant's eye (*adonis annua seu autumnalis* of Smith) fl.	July 1	Apr 11, July 15
Red eyebright (*euphrasia seu bartsia odontites* of Smith) fl.	July 2	June 20, Aug 10
Thorough wax (*bupleurum rotundifolia*) fl.	July 2	
Cockle (*agrostemma githago*) fl.	July 2	May 14, July 25
Ivy-leaved wild lettuce (*prenanthes muralis*) fl.	July 2	June 2, July 25
Feverfew (*matricaria seu pyrethrum parthenium* of Smith) fl.	July 2	June 19, July 24
Wall pepper (*sedum acre*) fl.	July 3	June 8, July 12
Privet (*ligustrum vulgare*) fl.	July 3	June 3, July 13
Common toadflax (*antirrhinum linaria*) fl.	July 3	June 21, Aug 3
Perennial wild fl.ax (*linum perenne*) fl.	July 4	Apr 21, July 6
Whortleberries, ripe (*vaccinium ulig*)	July 4–24	
Yellow base rocket (*reseda lutea*) fl.	July 5	July 19
Blue-bottle (*centaurea cyanus*) fl.	July 5	May 15, Oct 14
Dwarf carline thistle (*carduus acaulis*) fl.	July 5–12	June 30, Aug 4
Bull-rush, or cat's-tail (*typha latifolia*) fl.	July 6	June 29, July 21
Spiked willow-herb (*lythrum salicaria*) fl.	July 6	June 24, Aug 17
Black mullein (*verbascum niger*) fl.	July 6	
Chrysanthemum coronarium, fl.	July 6	May 28, July 28
Marigolds (*calendula officinalis*) fl.	July 6–9	Apr 20, July 16
Little field madder (*sherardia arvensis*) fl.	July 7	Jan 11, June 6

	WHITE	MARKWICK
Calamint (*melissa seu thymus calamintha* of Smith) fl.	July 7	July 21
Black horehound (*ballota nigra*) fl.	July 7	June 16, Sept 12
Wood betony (*betonica officinalis*) fl.	July 8–19	June 10, July 15
Round-leaved bell-flower (*campanula rotundifolia*) fl.	July 8	June 12, July 29
All-good (*chenopodium bonus henricus*) fl.	July 8	Apr 21, June 15
Wild-carrot (*daucus carota*) fl.	July 8	June 7, July 14
Indian cress (*epopaeolum majus*) fl.	July 8–20	June 11, July 25
Cat-mint (*nepata cataria*) fl.	July 9	
Cow-wheat (*melampyrum sylvaticum seu pratense* of Smith) fl.	July 9	May 2, June 22
Crosswort (*valantia cruciatia seu galium cruciatum* of Smith) fl.	July 9	Apr 10, May 28
Cranberries ripe	July 9–27	
Tufted vetch (*vicia cracca*) fl.	July10	May 31, July 8
Wood vetch (*vicia sylvat.*) fl.	July 10	
Little throatwort (*campanula glomerata*) fl.	July 11	July 28, Aug 18
Sheep's scabious (*jasione montana*) fl.	July 11	June 10, July 25
Pastinaca sylv. fl.	July 12	
White lily (*lilium candidum*) fl.	July 12	June 21, July 22
Hemlock (*conium maculatum*) fl.	July 13	June 4, July 20
Caucalus anthriscus, fl.	July 13	
Flying ants, ap.	July 13–Aug 11	Aug 29, Sept 19
Moneywort (*lysimachia nummularia*) fl.	July 13	June 14, Aug 16
Scarlet martagon (*lilium chalcedonicum*) fl.	July 14–Aug 4	June 21, Aug 6
Lesser stitchwort (*stillaria graminea*) fl.	July 14	May 8, June 23
Fool's parsley (*aethusa cynapium*) fl.	July 14	June 9, Aug 9
Dwarf elder (*sambucus ebulus*) fl.	July 14–29	
Swallows and martins congregate	July 14, Aug 29	Aug 12, Sept 8
Potato (*solanum tuberosum*) fl.	July 14	June 3, July 12
Angelica sylv. fl.	July 15	
Digitalis ferrugin, fl.	July 15–25	
Ragwort (*senecio jacobaea*) fl.	July 15	June 22, July 13
Golden rod (*solidago virgaurea*) fl.	July 15	July 7, Aug 29
Star thistle (*centaurea calc trapa*) fl.	July 16	July 16, Aug 16
Tree primrose (*oenothera biennis*) fl.	July 16	June 12, July 18
Peas (*pisum sativum*) cut	July 17, Aug 14	July 13, Aug 15

	WHITE	MARKWICK
Galega officina fl.	July 17	
Apricots (*prunus armeniaca*) ripe	July 17, Aug 21	July 5, Aug 16
Crown's allheal (*stachys palustris*) fl.	July 17	June 12, July 14
Branching willow-herb (*epilobium ramos*) fl.	July 17	
Rye-harvest begins	July 17, Aug 7	
Yellow centaury (*chlora perfoliata*) fl.	July 18, Aug 15	June 15, Aug 13
Yellow vetchling (*lathyrus aphaca*) fl.	July 18	
Enchanter's nightshade (*circaea lutetiana*) fl.	July 18	June 20, July 27
Water hemp agrimony (*eupatorium cannabinum*) fl.	July 18	July 4, Aug 6
Giant throatwort (*campanula trachelium*) fl.	July 19	July 13, Aug 14
Eyebright (*euphrasia officinalis*) fl.	July 19	May 28, July 19
Hops (*humulus lupulus*) fl.	July 19, Aug 10	July 20, Aug 17
Poultry moult	July 19	
Dodder (*cuscuta europaea seu epithymum* of Smith) fl.	July 20	July 9, Aug 7
Lesser centaury (*gentiana seu chironia centaurium* of Smith) fl.	July 20	June 3, July 19
Creeping water parsnep (*sium nodi florum*) fl.	July 20	July 10, Sept 11
Common spurrey (*spergula arvensis*) fl.	July 21	Apr 10, July 16
Wild clover (*trifolium pratense*) fl.	July 21	May 2, June 7
Buckwheat (*polygonum fagopyrum*) fl.	July 21	June 27, July 10
Wheat harvest begins	July 21, Aug 23	July 11, Aug 26
Great burr-reed (*sparganium erectum*) fl.	July 22	June 10, July 23
Marsh St John's-wort (*hypericum elodes*) fl.	July 22–31	June 16, Aug 10
Sun-dew (*drosera rotundifolia*) fl.	July 22	Aug 1
March cinquefoil (*comarum palustre*) fl.	July 22	May 27, July 12
Wild cherries ripe	July 22	
Lancashire asphodel (*anthericum ossi-fragum*) fl.	July 22	June 21, July 29
Hooded willow-herb (*scutellaria galericulata*) fl.	July 23	June 2, July 31
Water dropwort (*oenanthe fistulos*) fl.	July 23	
Horehound (*marrubium vulg.*) fl.	July 23	
Seseli caruifol. fl.	July 24	

	WHITE	MARKWICK
Water plantain (*alisma plantago*) fl.	July 24	May 1, July 31
Alopecurus myosuroides, fl.	July 25	
Virgin's bower (*clematis vitalba*) fl.	July 25, Aug 9	July 13, Aug 14
Bees kill the drones	July 25	
Teasel (*dipsacus sylvestris*) fl.	July 26	July 16, Aug 3
Wild marjoram (*origanum vulgare*) fl.	July 26	July 17, Aug 29
Swifts (*hirundo apus*) begin to depart	July 27–29	Aug 5
Small wild teasel (*dipsacus pilosus*) fl.	July 28, 29	
Wood sage (*teucrium scorodonia*) fl.	July 28	June 17, July 24
Everlasting pea (*lathyrus latifolius*) fl.	July 28	June 20, July 30
Trailing St John's-wort (*hypericum humifusum*) fl.	July 29	May 20, June 22
White hellebore (*veratrum album*) fl.	July 30	July 18–22
Camomile (*anthemis nobilis*) fl.	July 30	June 21, Aug 20
Lesser field scabius (*scabiosa columbaria*) fl.	July 30	July 13, Aug 9
Sunflower (*helianthus multiflorus*) fl.	July 31, Aug 6	July 4, Aug 22
Yellow loosestrife (*lysimachia vulgaris*) fl.	July 31	July 2, Aug 7
Swift (*hirundo apus*) last seen	July 31, Aug 27	Aug 11
Oats (*avena sativa*) cut	Aug 1–16	July 26, Aug 19
Barley (*hordeum sativum*) cut	Aug 1–26	July 27, Sept 4
Lesser hooded willow-herb (*scutellaria minor*) fl.	Aug 1	Aug 8, Sept 7
Middle fleabane (*inula disinterica*) fl.	Aug 2	July 7, Aug 3
Apis manicata, ap.	Aug 2	
Swallow-tailed butterfly (*papilio machaon*) ap.	Aug 2	Apr 20, June 7 last seen Aug 28
Whame or burrel-fly (*oestrus bovis*) lays eggs on horses	Aug 3–19	
Sow thistle (*sonchus arvensis*) fl.	Aug 3	June 17, July 21
Plantain fritillary (*papilio cinxia*) ap.	Aug 3	
Yellow succory (*picris hieracioides*) fl.	Aug 4	June 6–25
Musca mystacea, ap.		
Canterbury bells (*campanula medium*) fl.	Aug 5	June 5, Aug 11
Mentha longifol. fl.	Aug 5	
Carline thistle (*carlina vulgaris*) fl.	Aug 7	July 21, Aug 18
Venetian sumach (*rhus cotinus*) fl.	Aug 7	June 5, July 20
Ptinus pectinicornis, ap.	Aug 7	
Burdock (*arctium lappa*) fl.	Aug 8	June 17, Aug 4

	WHITE	MARKWICK
Fell-wort (*gentiana amarella*) fl.	Aug 8, Sept 3	
Wormwood (*artemisia absinthium*) fl.	Aug 8	July 22, Aug 21
Mugwort (*artemisia vulgaris*) fl.	Aug 8	July 9, Aug 10
St Barnaby's thistle (*centauria solstit*) fl.	Aug 10	
Meadow saffron (*colchicum autumnale*) fl.	Aug 10, Sept 13	Aug 15, Sept 29
Michaelmas daisy (*aster tradescantia*) fl.	Aug 12, Sept 27	Aug 11, Oct 8
Meadow rue (*thalictrum flavum*) fl.	Aug 14	
Sea holly (*eryngium marit.*) fl.	Aug 14	
China aster (*aster chinensis*) fl.	Aug 15, Sept 28	Aug 6, Oct 2
Boletus albus, ap.	Aug 14	May 10
Less Venus looking-glass (*campanula hybrida*) fl.	Aug 15	May 14
Carthamus tinctor fl.	Aug 15	
Goldfinch (*fringilla carduelis*) young broods, ap.	Aug 15	June 15
Lapwings (*tringa vanellus*) congregate	Aug 15, Sept 12	Sept 25, Feb 4
Black-eyed marble butterfly (*papilio semele*) ap.	Aug 15	
Birds reassume their spring notes	Aug 16	
Devil's bit (*scabiosa succisa*) fl.	Aug 17	June 22, Aug 23
Thistle-down floats	Aug 17, Sept 10	
Ploughman's spikenard (*conyza squarrosa*) fl.	Aug 18	
Autumnal dandelion (*leontodon autumnale*) fl.	Aug 18	July 25
Flies about in windows	Aug 18	
Linnets (*fringilla linota*) congregate	Aug 18, Nov 1	Aug 22, Nov 8
Bulls make their shrill autumnal noise	Aug 20	
Aster amellus, fl.		Aug 22
Balsam (*impatiens balsamina*) fl.	Aug 23	May 22, July 26
Milk thistle (*carduus marinus*) fl.	Aug 24	Apr 21, July 18
Hop-picking begins	Aug 24, Sept 17	Sept 1–15
Beech (*fagus sylvatica*) turns yellow	Aug 24, Sept 22	Sept 5–29
Soapwort (*saponaria officinalis*) fl.	Aug 25	July 19, Aug 23
Ladies' traces (*ophrys spiralis*) fl.	Aug 27, Sept 12	Aug 18, Sept 18
Small golden black-spotted butterfly (*papilio phlaeas*) ap.	Aug 29	
Swallow (*hirundo rustica*) sings	Aug 29	Apr 11, Aug 20
Althaea frutex (*hibiscus syriacus*) fl.	Aug 30, Sept 2	July 20, Sept 28

	WHITE	MARKWICK
Great fritillary (*papilio paphia*) ap.	Aug 30	
Willow red under-wing moth (*phalaena pacta*) ap.	Aug 31	
Stone curlew (*otis oedicnemus*) clamours	Sept 1, Nov 7	June 17
Phalaena russula, ap.	Sept 1	
Grapes ripen	Sept 4, Oct 24	Aug 31, Nov 4
Wood-owls hoot	Sept 4, Nov 9	
Saffron butterfly (*papilio hyale*) ap.	Sept 4	Aug 5, Sept 26
Ring-ousel appears on its autumnal visit	Sept 4–30	
Flycatcher (*muscicapa grisola*) last seen	Sept 6–29	Sept 4–30
Beans (*vicia faba*) cut	Sept 11	Aug 9, Oct 14
Ivy (*hedera helix*) fl.	Sept 12, Oct 2	Sept 18, Oct 28
Stares congregate	Sept 12, Nov 1	June 4, Mar 21
Wild honeysuckles fl. a second time	Sept 25	
Woodlark sings	Sept 28, Oct 24	
Woodcock (*scolopax rusticola*) returns	Sept 29, Nov 11	Oct 1, Nov 1 young ones last seen Apr 28
Strawberry-tree (*arbutus unedo*) fl.	Oct 1	May 21, Dec 10
Wheat sown	Oct 3, Nov 9	Sept 23, Oct 19
Swallows last seen (NB The house-martin the latest)	Oct 4, Nov 5	Nov 16
Redwing (*turdus iliacus*) comes	Oct 10, Nov 10	Oct 1, Dec 18 sings Feb 10, Mar 21 last seen April 13
Fieldfare (*turdus pilaris*) returns	Oct 12, Nov 23	Oct 13, Nov 19 last seen May 1
Gossamer fills the air	Oct 15–27	
Chinese holly-oak (*alcea rosea*) fl.	Oct 19	July 7, Aug 21
Hen chaffinches congregate	Oct 20, Dec 31	
Wood-pigeons come	Oct 23, Dec 27	
Royston Crow (*corvus cornix*) returns	Oct 23, Nov 29	Oct 13, Nov 1 last seen Apr15
Snipe (*scolopax gallinago*) returns	Oct 25, Nov 20	Sept 29, Nov 11 last seen Apr 14
Tortoise begins to bury himself	Oct 27, Nov 26	
Rooks (*corvus frugilegus*) return to their nest-trees	Oct 31, Dec 25	June 29, Oct 20
Bucks grunt	Nov 1	
Primrose (*primula vulgaris*) fl.	Nov 10	Oct 7, Dec 30

	WHITE	MARKWICK
Green whistling plover, ap.	Nov 13, 14	
Helvella mitra, ap.	Nov 16	
Greenfinches flock	Nov 27	
Hepatica, fl.	Nov 30, Dec 29	Feb 19
Furze (*ulex europaeus*) fl.	Dec 4–21	Dec 16–31
Polyanthus (*primula polyanthus*) fl.	Dec 7–16	Dec 31
Young lambs dropped	Dec 11–27	Dec 12, Feb 21
Moles work in throwing up hillocks	Dec 12–23	
Helleborus foetidus, fl.	Dec 14–30	
Daisy (*bellis perennis*) fl.	Dec 15	Dec 26–31
Wall-flower (*cheiranthus cheiri seu fruticulosus of Smith*) fl.	Dec 15	Nov 5
Mezereon, fl.	Dec 15	
Snowdrop, fl.	Dec 29	

In sese vertitur annus

In Selborne church

POEMS

reprinted from the edition of 1813

The Invitation to Selborne

See Selborne spreads her boldest beauties round,
The varied valley, and the mountain ground,
Wildly majestic! what is all the pride
Of flats, with loads of ornament supply'd
Unpleasing, tasteless, impotent expense,
Compared with Nature's rude magnificence.
  Arise, my stranger, to these wild scenes haste;
The unfinish'd farm awaits your forming taste:
Plan the pavilion, airy, light, and true;
Thro' the high arch call in the length'ning view;
Expand the forest sloping up the hill;
Swell to a lake the scant, penurious rill;
Extend the vista, raise the castle mound
In antique taste with turrets ivy-crown'd;
O'er the gay lawn the flow'ry shrub dispread,
Or with the blending garden mix the mead;
Bid China's pale, fantastic fence delight;
Or with the mimic statue trap the sight.
  Oft on some evening, sunny, soft, and still,
The Muse shall lead thee to the beech-grown hill,
To spend in tea the cool, refreshing hour,
Where nods in air the pensile, nest-like bower;[1]
Or where the Hermit hangs the straw-clad cell;[2]
Emerging gently from the leafy dell;
By Fancy plann'd; as once th' inventive maid
Met the hoar sage amid the secret shade;
Romantic spot! from whence in prosper lies
Whate'er of landscape charms our feasting eyes;
The pointed spire, the hall, the pasture-plain,

1 A kind of arbour on the side of a hill.

2 A grotesque building, contrived by a young gentleman, who used on occasion to appear in the character of a hermit.

The russet fallow, or the golden grain,
The breezy lake that sheds a gleaming light,
Till all the fading picture fail the sight.
Each to his task; all different ways retire;
Cull the dry stick; call forth the seeds of fire;
Deep fix the kettle's props, a forky row,
Or give with fanning hat the breeze to blow.
Whence is this taste, the furnish'd hall forgot,
To feast in gardens, or the unhandy grot?
Or novelty with some new charms surprises,
Or from our very shifts some joy arises.
Hark, while below the village bells ring round,
Echo, sweet nymph, returns the soften'd sound;
But if gusts rise, the rushing forests roar,
Like the tide tumbling on the pebbly shore.
Adown the vale, in lone, sequester'd nook,
Where skirting woods embrown the dimpling brook,
The ruin'd Convent lies; here wont to dwell
The lazy canon midst his cloistr'd cell;[3]
While papal darkness brooded o'er the land,
Ere Reformation made her glorious stand:
Still oft at eve belated shepherd-swains
See the cowl'd spectre skim the folded plains.
To the high Temple would my stranger go,[4]
The mountain-brow commands the woods below;
In Jewry first this order found a name,
When madding Croisades set the world in flame;
When western climes, urg'd on by Pope and priest,
Pour'd forth their millions o'er the deluged East:
Luxurious knights, ill suited to defy
To mortal fight Turcéstan chivalry.
Nor be the Parsonage by the muse forgot;
The partial bard admires his native spot;
Smit with its beauties, loved, as yet a child,
(Unconscious why) its scapes grotesque, and wild.
High on a mound th' exalted gardens stand.

3 The ruins of a priory, founded by Peter de Rupibus, Bishop of Winchester.
4 The remains of a preceptory of the Knights Templar; at least it was a farm dependent upon some preceptory of that order. I find it was a preceptory, called the Preceptory of Sudington; now called Southington.

Beneath, deep valleys scoop'd by Nature's hand.
A Cobham here, exulting in his art,
Might blend the General's with the Gardener's part;
Might fortify with all the martial trade
Of rampart, bastion, fosse, and palisade;
Might plant the mortar with wide threat'ning bore,
Or bid the mimic canon seem to roar.
  Now climb the steep, drop now your eye below,
Where round the blooming village orchards grow;
There, like a picture, lies my lowly seat,
A rural, shelter'd, unobserv'd retreat.
  Me far above the rest Selbornian scenes,
The pendent forest, and the mountain-greens
Strike with delight; there spreads the distant view,
That gradual fades till sunk in misty blue:
Here Nature hangs her slopy woods to sight,
Rills purl between, and dart a quivering light.

Selborne Hanger – A Winter Piece

To the Miss Batties

The Bard, who sang so late in blithest strain
Selbornian prospects, and the rural reign,
Now suits his plaintive pipe to sadden'd tone,
While the blank swains the changeful year bemoan.
  How fall'n the glories of these fading scenes!
The dusky beech resigns his vernal greens,
The yellow maple mourns in sickly hue,
And russet woodlands crowd the dark'ning view.
  Dim, clust'ring fogs involve the country round,
The valley and the blended mountain ground
Sink in confusion; but with tempest-wing
Should Boreas from his northern barrier spring,
The rushing woods with deaf'ning clamour roar,
Like the sea tumbling on the pebbly shore.
When spouting rains descend in torrent tides,
See the torn Zigzag weep its channel'd sides:
Winter exerts its rage: heavy and slow,

From the keen east rolls on the treasur'd snow;
Sunk with its weight the bending boughs are seen,
And one bright deluge whelms the works of men.
Amidst this savage landscape, bleak and bare,
Hangs the chill hermitage in middle air;
Its haunts forsaken, and its feasts forgot,
A leaf-strown, lonely, desolated cot!
  Is this the scene that late with rapture rang,
Where Delphy danc'd, and gentle Anna sang;
With fairy-step where Harriet tripp'd so late,
And on her stump reclined the musing Kitty sate?
  Return, dear nymphs; prevent the purple spring,
Ere the soft nightingale essays to sing;
Ere the first swallow sweeps the fresh'ning plain,
Ere love-sick turtles breathe their amorous pain
Let festive glee th' enliven'd village raise,
Pan's blameless dance the smitten swain surprise,
And bring all Arcady before our eyes.
  Return, blithe maidens; with you bring along
Free, native humour, all the charms of song,
The feeling heart, and unaffected ease,
Each nameless grace, and ev'ry power to please.

November 1, 1763

On the Rainbow

Look upon the rainbow, and praise him that made it; very beautiful is the brightness thereof.

ECCLESIASTES XLIII: 11

On morning or on ev'ning cloud impress'd,
Bent in vast curve, the watery meteor shines
Delightfully, to th' levell'd sun oppos'd:
Lovely refraction! while the vivid brede
In listed colours glows, th' unconscious swain
With vacant eye gazes on the divine
Phenomenon, gleaming o'er the illumined fields,
Or runs to the treasures which it sheds.
  Not so the sage, inspir'd with pious awe;

He hails the federal arch;[5] and looking up,
Adores that God, whose fingers form'd this bow
Magnificent, compassing heav'n about,
With a resplendent verge, 'Thou madest the cloud,
Maker omnipotent, and thou the bow;
Any by that covenant graciously hast sworn
Never to drown the world again;[6] henceforth,
Till time shall be no more, in ceaseless round,
Season shall follow season: day to night,
Summer to winter, harvest to seed time,
Heat shall to cold in regular array
Succeed.' - Heaven-taught, so sang the Hebrew bard.[7]

A Harvest Scene

Wak'd by the gentle gleamings of the morn,
Soon glad, the reaper, provident of want,
Hies cheerful-hearted to the ripen'd field;
Nor hastes alone; attendant by his side
His faithful wife, sole partner of his cares,
Bears on her breast the sleeping babe; behind,
With steps unequal, trips her infant train;
Thrice happy pair, in love and labour join'd!
All day they ply their task; with mutual chat,
Beguiling each the sultry, tedious hours.
Around them falls in rows the sever'd corn.
Or the shocks rise in regular array.
But when high noon invites to short repast,
Beneath the shade of shelt'ring thorn they sit,
Divide the simple meal, and drain the cask:
The swinging cradle lulls the whimp'ring babe,
Meantime; while growling round, if at the tread
Of hasty passenger alarm'd, as of their store
Protective, stalks the cur with bristling back,
To guard the scanty scrip and russet frock.

5 Genesis IX: 12–17 6 Genesis viii: 22 7 Moses

On the Early and Late Blowing of the Vernal and Autumnal Crocus

Say, what impels amidst surrounding snow
Congeal'd the Crocus' flamy bud to grow;
Say, what retards amidst the summer's blaze
Th' autumnal bulb; till pale declining days?
The GOD of SEASONS! whose pervading power
Controls the sun, or sheds the fleecy shower;
He bids each flower his quick'ning word obey,
Or to each lingering bloom enjoins delay.

On the Dark, Still, Dry, Warm Weather, Occasionally Happening in the Winter Months

Th' imprison'd winds slumber within their caves
Fast bound: the fickle vane, emblem of change,
Wavers no more, long settling to a point.
All nature nodding seems composed: thick streams
From land, from flood up-drawn, dimming the day
'Like a dark ceiling stand': slow thro' the air
Gossamer floats, or stretch'd from blade to blade
The wavy net-work whitens all the field.
Push'd by the weightier atmosphere, up springs
The ponderous Mercury, from scale to scale
Mounting, amidst the Torricellian tube.[8]
While high in air, and pois'd upon his wings
Unseen, the soft, enamour'd wood-lark runs
Thro' all his maze of melody; – the brake
Loud with the black-bird's bolder note resounds.
Sooth'd by the genial warmth, the cawing rook
Anticipates the spring, selects her mate,

8 The Barometer.

Haunts her tall nest-trees, and with sedulous care
Repairs her wicker eyrie, tempest torn.
   The ploughman inly smiles to see upturn
His mellow glebe, best pledge of future crop:
With glee the gardener eyes his smoking beds:
E'en pining sickness feels a short relief.
   The happy schoolboy brings transported forth
His long forgotten scourge, and giddy gig:
O'er the white paths he whirls the rolling hoop,
Or triumphs in the dusty fields of taw.
   Not so the museful sage: – abroad he walks
Contemplative, if haply he may find
What cause controls the tempest's rage, or whence
Amidst the savage season winter smiles.
   For days, for weeks, prevails the placid calm.
At length some drops prelude a change: the sun
With ray refracted bursts the parting gloom;
When all the chequer'd sky is one bright glare.
   Mutters the wind at eve: th' horizon round
With angry aspect scowls: down rush the showers,
And float the delug'd paths, and miry fields.

WORDSWORTH CLASSICS

The Mill on the Floss
Middlemarch
Silas Marner

HENRY FIELDING
Tom Jones

RONALD FIRBANK
Valmouth & other stories

TRANSLATED BY
EDWARD FITZGERALD
The Rubaiyat of Omar Khayyam

F. SCOTT FITZGERALD
The Diamond as Big as the Ritz & other stories
The Great Gatsby
Tender is the Night

GUSTAVE FLAUBERT
Madame Bovary

JOHN GALSWORTHY
In Chancery
The Man of Property
To Let

ELIZABETH GASKELL
Cranford
North and South

GEORGE GISSING
New Grub Street

KENNETH GRAHAME
The Wind in the Willows

GEORGE & WEEDON GROSSMITH
Diary of a Nobody

RIDER HAGGARD
She

THOMAS HARDY
Far from the Madding Crowd
Jude the Obscure
The Mayor of Casterbridge
A Pair of Blue Eyes
The Return of the Native
Selected Short Stories
Tess of the D'Urbervilles
The Trumpet Major
Under the Greenwood Tree
Wessex Tales
The Woodlanders

NATHANIEL HAWTHORNE
The Scarlet Letter

O. HENRY
Selected Stories

JAMES HOGG
The Private Memoirs and Confessions of a Justified Sinner

HOMER
The Iliad
The Odyssey

E. W. HORNUNG
Raffles: The Amateur Cracksman

VICTOR HUGO
The Hunchback of Notre Dame
Les Misérables
IN TWO VOLUMES

HENRY JAMES
The Ambassadors
Daisy Miller & other stories
The Europeans
The Golden Bowl
The Portrait of a Lady
The Turn of the Screw & The Aspern Papers

M. R. JAMES
Ghost Stories

JEROME K. JEROME
Three Men in a Boat

JAMES JOYCE
Dubliners
A Portrait of the Artist as a Young Man

RUDYARD KIPLING
The Best Short Stories
Captains Courageous
Kim
The Man Who Would Be King & other stories
Plain Tales from the Hills

D. H. LAWRENCE
Lady Chatterley's Lover
The Plumed Serpent
The Rainbow
Sons and Lovers
The Virgin and the Gypsy & selected stories
Women in Love

SHERIDAN LE FANU
(EDITED BY M. R. JAMES)
Madam Crowl's Ghost & other stories
In a Glass Darkly

GASTON LEROUX
The Phantom of the Opera

JACK LONDON
Call of the Wild & White Fang

GUY DE MAUPASSANT
The Best Short Stories

HERMAN MELVILLE
Moby Dick
Typee

GEORGE MEREDITH
The Egoist

H. H. MUNRO
The Complete Stories of Saki

WORDSWORTH CLASSICS

THOMAS LOVE PEACOCK
Headlong Hall & Nightmare Abbey

EDGAR ALLAN POE
Tales of Mystery and Imagination

FREDERICK ROLFE
Hadrian the Seventh

SIR WALTER SCOTT
Ivanhoe
Rob Roy

WILLIAM SHAKESPEARE
All's Well that Ends Well
Antony and Cleopatra
As You Like It
The Comedy of Errors
Coriolanus
Hamlet
Henry IV Part 1
Henry IV Part 2
Henry V
Julius Caesar
King John
King Lear
Love's Labours Lost
Macbeth
Measure for Measure
The Merchant of Venice
The Merry Wives of Windsor
A Midsummer Night's Dream
Much Ado About Nothing
Othello
Pericles
Richard II
Richard III
Romeo and Juliet
The Taming of the Shrew
The Tempest
Titus Andronicus
Troilus and Cressida
Twelfth Night
Two Gentlemen of Verona
A Winter's Tale

MARY SHELLEY
Frankenstein

TOBIAS SMOLLETT
Humphry Clinker

LAURENCE STERNE
A Sentimental Journey
Tristram Shandy

ROBERT LOUIS STEVENSON
Dr Jekyll and Mr Hyde
The Master of Ballantrae & Weir of Hermiston

BRAM STOKER
Dracula

R. S. SURTEES
Mr Sponge's Sporting Tour

JONATHAN SWIFT
Gulliver's Travels

W. M. THACKERAY
Vanity Fair

TOLSTOY
Anna Karenina
War and Peace

ANTHONY TROLLOPE
Barchester Towers
Can You Forgive Her?
Dr Thorne
The Eustace Diamonds
Framley Parsonage
The Last Chronicle of Barset
Phineas Finn
Phineas Redux
The Small House at Allington
The Way We Live Now
The Warden

IVAN SERGEYEVICH TURGENEV
Fathers and Sons

MARK TWAIN
Tom Sawyer & Huckleberry Finn

JULES VERNE
Around the World in Eighty Days & Five Weeks in a Balloon
Journey to the Centre of the Earth
Twenty Thousand Leagues Under the Sea

VIRGIL
The Aeneid

VOLTAIRE
Candide

LEW WALLACE
Ben Hur

ISAAC WALTON
The Compleat Angler

EDITH WHARTON
The Age of Innocence

GILBERT WHITE
The Natural History of Selborne

OSCAR WILDE
Lord Arthur Savile's Crime & other stories
The Picture of Dorian Gray
The Plays
IN TWO VOLUMES

VIRGINIA WOOLF
Mrs Dalloway
Orlando
To the Lighthouse

P. C. WREN
Beau Geste

Bhagavad Gita

Distribution

AUSTRALIA & PAPUA NEW GUINEA
Peribo Pty Ltd
58 Beaumont Road, Mount Kuring-Gai
NSW 2080, Australia
Tel: (02) 457 0011 Fax: (02) 457 0022

CZECH REPUBLIC
Bohemian Ventures s r. o.,
Delnicka 13, 170 00 Prague 7
Tel: 042 2 877837 Fax: 042 2 801498

FRANCE
Copernicus Diffusion
23 Rue Saint Dominique, Paris 75007
Tel: 1 44 11 33 20 Fax: 1 44 11 33 21

GERMANY & AUSTRIA
GLBmbH (Bargain, Promotional & Remainder Shops)
Zollstockgürtel 5, 50969 Köln
Tel: 0221 34 20 92 Fax: 0221 38 40 40

Tradis Verlag und Vertrieb GmbH (Bookshops)
Postfach 90 03 69, D-51113 Köln
Tel: 022 03 31059 Fax: 022 03 3 93 40

GREAT BRITAIN & IRELAND
Wordsworth Editions Ltd
Cumberland House, Crib Street
Ware, Hertfordshire SG12 9ET

INDIA
OM Book Service
1690 First Floor, Nai Sarak, Delhi – 110006
Tel: 3279823-3265303 Fax: 3278091

ISRAEL
Timmy Marketing Limited
Israel Ben Zeev 12, Ramont Gimmel, Jerusalem
Tel: 02-865266 Fax: 02-880035

ITALY
Magis Books s.p.a.
Via Raffaello 31/C, Zona Ind Mancasale
42100 Reggio Emilia
Tel: 0522 920999 Fax: 0522 920666

NEW ZEALAND & FIJI
Allphy Book Distributors Ltd
4-6 Charles Street, Eden Terrace, Auckland,
Tel: (09) 3773096 Fax: (09) 3022770

MALAYSIA & BRUNEI
Vintrade SDN BHD
5 & 7 Lorong Datuk Sulaiman 7
Taman Tun Dr Ismail
60000 Kuala Lumpur, Malaysia
Tel: (603) 717 3333 Fax: (603) 719 2942

MALTA & GOZO
Agius & Agius Ltd
42A South Street, Valletta VLT 11
Tel: 234038 - 220347 Fax: 241175

NORTH AMERICA
Universal Sales & Marketing
230 Fifth Avenue, Suite 1212
New York, NY 10001, USA
Tel: 212 481 3500 Fax: 212 481 3534

PHILIPPINES
I J Sagun Enterprises
P O Box 4322 CPO Manila
2 Topaz Road, Greenheights Village,
Taytay, Rizal
Tel: 631 80 61 TO 66

PORTUGAL
International Publishing Services Ltd
Rua da Cruz da Carreira, 4B, 1100 Lisbon
Tel: 01 570051 Fax: 01 3522066

SOUTHERN & CENTRAL AFRICA
Southern Book Publishers (Pty) Ltd
P.O.Box 3103
Halfway House 1685, South Africa
Tel: (011) 315-3633/4/5/6
Fax: (011) 315-3810

EAST AFRICA & KENYA
P.M.C. International Importers & Exporters CC
Unit 6, Ben-Sarah Place, 52-56 Columbine Place,
Glen Anil, Kwa-Zulu Natal 4051,
P.O.Box 201520,
Durban North, Kwa-Zulu Natal 4016
Tel: (031) 844441 Fax: (031) 844466

SINGAPORE
Paul & Elizabeth Book Services Pte Ltd
163 Tanglin Road No 03-15/16
Tanglin Mall, Singapore 1024
Tel: (65) 735 7308 Fax: (65) 735 9747

SLOVAK REPUBLIC
Slovak Ventures s r. o.,
Stefanikova 128, 949 01 Nitra
Tel/Fax: 042 87 525105/6/7

SPAIN
Ribera Libros, S.L.
Poligono Martiartu, Calle 1 - no 6
48480 Arrigorriaga, Vizcaya
Tel: 34 4 6713607 (Almacen)
34 4 4418787 (Libreria)
Fax: 34 4 6713608 (Almacen)
34 4 4418029 (Libreria)

UNITED ARAB EMIRATES
Nadoo Trading LLC
P.O.Box 3186
Dubai
United Arab Emirates
Tel: 04-359793 Fax: 04-487157

DIRECT MAIL **Bibliophile Books**
5 Thomas Road, London E14 7BN,
Tel: 0171-515 9222 Fax: 0171-538 4115
Order hotline 24 hours Tel: 0171-515 9555
Cash with order + £2.00 p&p (UK)